Climate Change

A Holistic View

Climate Change

A Holistic View

R. R. Kelkar

M. Sc., Ph. D.

Former Director General of Meteorology
India Meteorological Department
and
Former ISRO Space Chair Professor
University of Pune

BSP BS Publications

A unit of **BSP Books Pvt. Ltd.**

4-4-309/316, Giriraj Lane, Sultan Bazar,
Hyderabad - 500 095
Phone : 040 - 23445605, 23445688

Published by :

BSP **BS Publications**
A unit of **BSP Books Pvt. Ltd.**

4-4-309/316, Giriraj Lane, Sultan Bazar,
Hyderabad - 500 095
Phone : 040 - 23445605, 23445688
e-mail : info@bspbooks.net

ISBN : 978-93-52300-55-6 (HB)

This book is dedicated

to the memory of my mother

INDUMATI RATNAKAR KELKAR

Preface

In the last few years, I have had the pleasure and privilege of speaking about weather, climate and climate change on more than fifty occasions at many different places. The invitations came from research institutes, professional societies, educational institutions, management schools, training courses, workshops, conferences and brainstorming sessions. They gave me the opportunity to address not only meteorologists but geologists, geographers, agricultural scientists, industrialists, farmers, teachers, schoolchildren, college students, journalists, and even senior citizens' groups. My lectures covered various topics like the science of climate change, satellite-based climate monitoring, extreme events, natural disasters, tropical cyclones, climate of the holocene, impacts of global warming on India, future climates, and others.

During the interactions with my diverse audiences, I often sensed that it was not easy for people to grasp the range of complexities of climate change. Some focussed only on the science part, many were over-concerned about the impacts, others wanted to generate awareness among common people. Climate change has many aspects, but they have to be understood and considered in their totality and not in isolation. The motivation for writing this book came from the need to present a holistic view of climate change with special reference to India.

The science of climate change considers the observational evidence, attempts to model the processes within the climate system, and ventures to make projections and predictions of climate over the next century or longer. Climate change has a major political angle, because when developed and developing countries come together to discuss it, the past history of exploitation comes into conflict with the hopes for a better future. Economics also comes into the picture as unabated climate change is sure to harm national economies, while mitigation actions have their own costs. There are ethical issues as well, as the entire human society will be affected by climate change, and there has to be a deep consideration of whether our actions are right or wrong, just or unjust, good or bad.

This book has attempted to take a rational and balanced view of climate change. First of all, it explains that the science of climate change is difficult because it has to deal with variables that have small magnitudes and large errors and uncertainties. For example, the energy received by the earth from the sun is 1366 watts/metre^2 while the anthropogenic radiative forcing is of the order of 1-2 watts/metre^2. Warm years are ranked in terms of global average temperature anomalies of 0.01 °C, but the value of the annual average surface temperature of the earth is itself uncertain and it could be around 13.9 °C or 14 °C, depending upon how we do the averaging. Global warming could result in a sea level rise of the order of a couple of millimeters per year with an uncertainty of half a millimeter, in an ocean couple of kilometres deep. One of the objectives of this book is to make people aware of the relative magnitudes of climate parameters, the changes that they are undergoing and the uncertainties in our assessments, so that we can put them in the proper perspective and not paint scary scenarios of the future.

The second point that this book makes is that India's contribution to past warming of the earth has been minimal compared to the developed and industrialized countries, and its future greenhouse gas emissions are also expected to be remain much lower in comparison. India is the victim of global warming and not its cause and there is no need to develop a sense of guilt. Climate change is not the only problem that India is facing. There are other equally compelling issues like poverty, illiteracy, health and food security. However, we have a responsibility towards future generations and we must do whatever we can to mitigate the effects of climate change while following the path of sustainable development. Going back to the dark ages is not the solution for global warming.

Pune
April 2010

R. R. Kelkar
kelkar_rr@yahoo.com

Acknowledgements

While writing this book on climate change, I have consulted a large number of papers and reports on the subject and visited several web sites on the internet that have related information. Wherever published results and methodologies are discussed in this book, the authors' names have been mentioned and the publication details are given in the list of references at the end of each chapter. For the many results, figures and tables that have been reproduced from the Fourth Assessment Report of the Intergovernmental Panel on Climate Change (IPCC), the source has been duly acknowledged and the authors' and editors' names are given in the references.

Where facts and data have been quoted from internet sources in the public domain, the web site addresses have been given. Figures 4.5.2.1 and 4.5.5.1 are reproduced from the web site of the University of Colorado at Boulder with its kind permission.

I must express my sincere thanks to Dr Ajit Tyagi, Director General, India Meteorological Department for the many figures, tables, data and imagery that I have included in this book. IMD has been cited as the source in the appropriate places. I am particularly grateful to Dr H. R. Hatwar, Dr A. K. Sharma, Dr Medha Khole, Dr D. S. Pai, Dr O. P. Singh and Dr M. Rajeevan, for their help in this regard.

Finally, I wish to express my sincere appreciation to BS Publications, Hyderabad, and particularly to Mr Anil Shah and Mr K. S. Raju, for the timely publication of this book and its very elegant printing.

R. R. Kelkar

Contents

Acronyms and Abbreviations

AAU	Assigned Amount Unit
ACE	Accumulated Cyclone Energy
AGCM	Atmospheric General Circulation Model
AOGCM	Atmosphere Ocean General Circulation Model
AR4	Fourth Assessment Report of IPCC
BAPMON	Background Pollution Monitoring Network
CAPE	Convective Available Potential Energy
CCN	Cloud Condensation Nuclei
CDM	Clean Development Mechanism
CER	Certified Emission Reduction
CFC	Chloro-Fluoro-Carbon
CINE	Convective Inhibition Energy
DMETER	Development of a European Multi-model Ensemble
DSSAT	Decision Support System for Agro-technology Transfer
EMIC	Earth System Model of Intermediate Complexity
ENSO	El Nino Southern Oscillation
ERU	Emission Reduction Unit
GARP	Global Atmospheric Research Programme,
GAW	Global Atmosphere Watch
GCN	Global Core Network
GCOS	Global Climate Observing System
GDP	Gross Domestic Product
GHG	GreenHouse Gas
GIA	Global Isostatic Anomaly
GISS	Goddard Institute for Space Studies
GIS	Geographical Information System
GLOSS	Global Sea Level Observing System
GO3OS	Global Ozone Observing System
GOME	Global Ozone Monitoring Experiment
GOOS	Global Ocean Observing System
GOS	Global Observing System

GPCP	Global Precipitation Climatology Project
GRACE	Gravity Recovery and Climate Experiment
GSI	Geological Survey of India
GTOS	Global Terrestrial Observing System
HadCRUT	U. K. Met Office Hadley Centre and
	Climatic Research Unit of the University of East Anglia
HCFC	Hydro-Chloro-Fluoro-Carbon
ICESAT	Ice Cloud and Land Elevation Satellite
ICSU	International Council for Science
IGY	International Geophysical Year
IMD	India Meteorological Department
IOC	Intergovernmental Oceanographic Commission
IPCC	Intergovernmental Panel on Climate Change
ISCCP	International Satellite Cloud Climatology Project
IWTC	International Workshop on Tropical Cyclones
JMA	Japan Meteorological Agency
LULUCF	Land Use, Land Use Change and Forestry
MMD	Multi-Model Data
NASA	U. S. National Aeronautics and Space Administration
NCDC	National Climatic Data Center
NDACC	Network for Detection of Atmospheric Composition Change
NOAA	U. S. National Oceanic and Atmospheric Administration
OCM	Ocean Circulation Model
ODS	Ozone Depleting Substances
OLR	Outgoing Longwave Radiation
OMI	Ozone Monitoring Instrument
OSTM	Ocean Surface Topography Mission
PDI	Power Dissipation Intensity
QBO	Quasi-Biennial Oscillation
QuikSCAT	Quick Scatterometer
RBCN	Regional Basic Climatological Network
RBSN	Regional Basic Synoptic Network
SBU	Solar Backscatter Ultraviolet
SCM	Simple Climate Model
SHADOZ	Southern Hemisphere Additional Ozonesondes

SMAP	Soil Moisture Active and Passive
SORCE	Solar Radiation and Climate Experiment
SRES	Special Report on Emission Scenarios
SST	Sea Surface Temperature
TAR	Third Assessment Report of IPCC
TERI	The Energy and Resources Institute
TOMS	Total Ozone Mapping Spectrometer
TRMM	Tropical Rainfall Measuring Mission
UNEP	United Nations Environment Programme
UNFCCC	United Nations Framework Convention on Climate Change
VOF	Voluntary Observing Fleet
WCP	World Climate Programme
WCRP	World Climate Research Programme
WDCGG	World Data Centre for Greenhouse Gases
WG I/II/III	Working Group I/II/III of IPCC
WMO	World Meteorological Organization
WTGROWS	Wheat Growth Simulator
WWW	World Weather Watch

Chapter 1

Evidence of Climate Change

Until the middle of the twentieth century, 'weather' and 'climate' had clear distinctions. Weather was the state of the atmosphere that existed at any given place and time, and it could be observed or measured with suitable instruments. Climate was the long-term average of the weather at a place, as shaped by various factors like its latitude, elevation or distance from the ocean. Different climates prevailed over different parts of the world and they could be classified into groups, the earliest such classification having been made by Koppen (1900) and later by Trewartha (1943). However, Thornthwaite (1948) was the first to call attention to the fact that the sum of climate elements that were observable did not necessarily amount to the climate, and subsequently Trewartha (1961) pointed out that many regions of the world had what he called problem climates.

Since weather and climate had their prescribed domains, meteorology and climatology developed and evolved as if they were two separate scientific entities. Meteorology was a physical science that dealt with the properties and processes of the atmosphere. Climatology was a descriptive science based upon the long-term means of weather observations, The derivation of climate normals was essentially a statistical exercise that smoothed out extreme values, but they served well as benchmarks against which the actual weather could be compared. Simply put, climate was what you expected to have, and weather was what you actually got.

However, outside of climatology was a school of thought that looked upon climate not as an invariant mean but as a variable governed by many factors, terrestrial as well as extra-terrestrial. It was also evident that on a geological time scale, the earth's climate had undergone significant changes and that the earth had experienced the so-called ice ages many times during its history, with intervening warm epochs. Some of the possible underlying causes could have been the changes in the earth's orbit or the inclination of its spin axis, variations in solar radiation and sunspot activity, and so on. Variation in the carbon dioxide content of the earth's atmosphere had also been thought of as a possible driving force behind the climate changes in the past. In the late 19th century, the general scientific opinion (Arrhenius 1896, Chamberlain 1899) was that carbon dioxide was indeed the prime cause of climate change. This hypothesis, however, received a setback in the early 20th century when it became known that the importance of water vapour in the radiative balance of the earth-atmosphere system was much greater than that of carbon dioxide (CO_2). The controversy was revived again when Callendar (1938, 1939) gave an estimate that a 30 % rise in the CO_2 content of the atmosphere would result in an increase of 1.1 °C in the global average surface temperature. A complete re-evaluation of the problem was made by Plass (1956) who concluded that the surface temperature will rise by 3.6 °C if the CO_2 concentration were to be doubled. Manabe et al (1967) were the first to consider the radiative equilibrium of the earth-atmosphere system in its entirety and they found that a doubling of the CO_2 in the atmosphere would lead to an increase of only 1.33 °C in the earth's mean surface temperature.

Kelkar (1970) had carried out an extensive investigation of the radiative equilibrium of the atmosphere over the Indian region using appropriate model atmospheres and making sensitivity experiments by changing the parameters of the model atmosphere. One of his simulation exercises had shown that an 100 % increase in the earth's CO_2 concentration would lead to an increase of 0.9 °C in the radiative equilibrium temperature of the earth's surface. Like in many similar studies made at that time, atmospheric dynamics or ocean processes had not been taken into consideration.

1.1 Climate and Climate Change

Until the middle of the 20th century, climate change was not talked about the way we all do today. Since then, our traditional perception of climate as a stable mean state of the atmosphere has itself been significantly altered, and climate variability as well as the magnitude and frequency of extreme events have come to be regarded as equally important. It is now recognized that climate change could not only be the result of natural causes but also be induced by human activities. The climate of the earth is now considered to be the net product of the interactions of land, vegetation, atmosphere and ocean, of physical, dynamical, chemical and biological processes, of carbon, nitrogen and water cycles. Human interventions in many of these processes are now occurring on a scale so unprecedented in history, that they are capable of bringing about climate change in a time span of decades to a century rather than over millennia as in the past.

Climate normals: If climate is regarded as average weather, then the period of averaging has to be given due consideration. The World Meteorological Organization (WMO) specifies successive 30-year periods such as 1931-1960, 1961-1990, etc, over which the normals of temperature, rainfall, pressure, wind and such other parameters should be computed by all national meteorological services. This helps to bring in uniformity and compatibility among data sets of various countries and in preparation of global charts and analyses. It also means that the climate normals are required to be updated every 30 years, using new data as it becomes available. However, many countries have long historical data sets and national practices differ with respect to the period of averaging. For example, India has prepared 50-year and 70-year rainfall normals as well.

Climate system: In a broader sense, the earth's climate system is a complex, interactive system consisting of five components:

1) atmosphere, or the envelope of gases around the earth, including clouds and aerosols,
2) land surface, of which land use and land use change are important aspects,
3) hydrosphere, comprising the oceans and all water bodies like rivers and lakes,
4) cryosphere, or the earth's snow and ice cover, including glaciers and ice sheets,
5) biosphere, consisting of all ecosystems and living organisms in the atmosphere, on land and in the ocean.

The five components of the earth's climate system also interact with each other (Figure 1.1.1). Thus there are earth-atmosphere or ocean-atmosphere interactions, or biogeochemical processes, taking place within the climate system. The climate system evolves in time under the influence of its own internal dynamics and external forcings, which could be natural phenomena like volcanic eruptions and solar variations, or anthropogenic changes in atmospheric composition. As the sun is the main source of energy that drives the earth's climate system, changes in the incoming solar radiation can modulate the climate system. Long term changes in the earth's orbital parameters, albedo, cloud cover, atmospheric aerosols, vegetation cover and concentrations of greenhouse gases can disturb the radiation balance of the earth's atmosphere. The climate system has feedback mechanisms which serve to restore the balance or bring the system to a new equilibrium state.

Climate change: The term climate change refers to a change in the mean state or its associated variability that persists for an extended period like decades or longer. The change should be identifiable, say by applying statistical tests. Climate change may result from either internal processes or external forcings or both. Some external influences may occur naturally, such as changes in solar radiation and volcanic activity. These may also contribute to the total natural variability of the climate system.

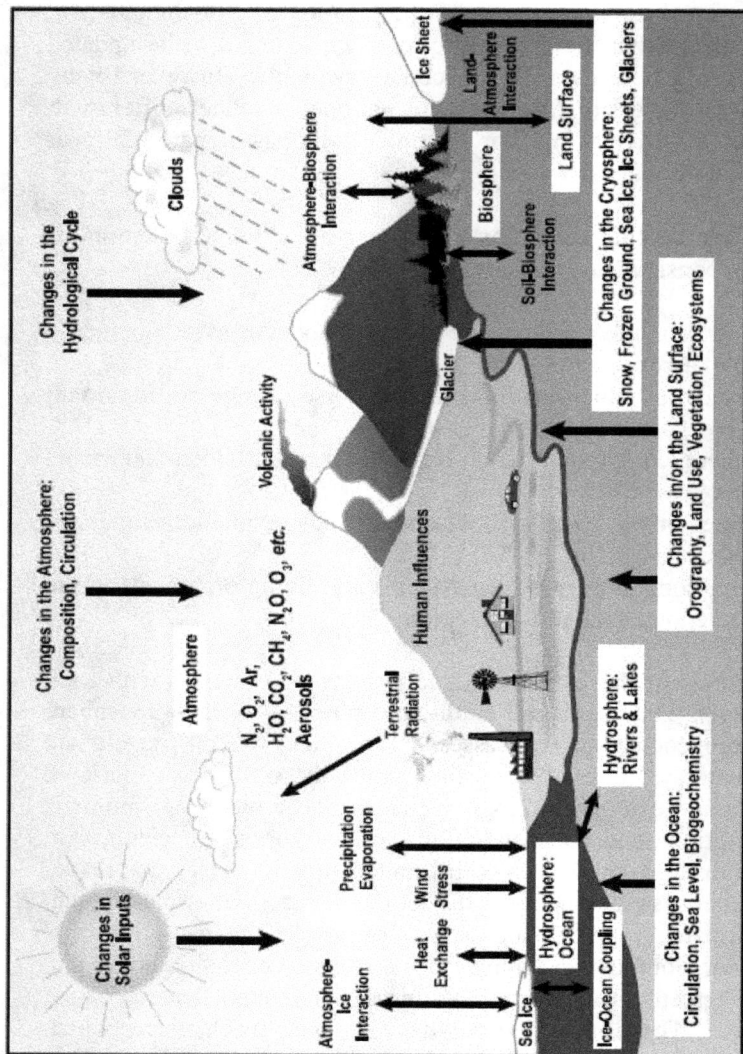

Figure 1.1.1 Schematic view of the components of the earth's climate system, important processes and their interactions. (Soirce: IPCC AR4, Le Treut et al 2007)

The Intergovernmental Panel on Climate Change (IPCC) defines climate change as any change in climate over time, whether due to natural variability or as a result of human activity. The United Nations Framework Convention on Climate Change (UNFCCC) defines climate change as a change of climate that is attributed directly or indirectly to human activity that alters the composition of the global atmosphere and that is in addition to natural climate variability.

Climate variability: This refers to the variations in the mean state of the climate and the statistics associated with the mean like standard atmosphere and extreme values, on all temporal and spatial scales beyond the scale of weather events. Climate variability could be the result of natural internal processes within the climate system or external forcings, both natural and anthropogenic.

1.2 Climate Issues in Today's World

Today's world is very different from what it was in the 19th century or in the middle of the 20th century in many ways, but particularly so with regard to our perceptions of climate. Currently, thousands of scientists around the globe are engaged in the study of climate, unlike in the past when it was the subject of speculative research by just a few. Climate change has become a household word. School students are being taught the importance of conserving and protecting the environment. Industrial projects are subjected to an environmental audit. National governments are more than willing to come together to discuss the state of the earth's environment and take concrete actions. Human society worldwide is aware that its very existence is threatened by global warming and climate change.

There were several individual scientific, technological and political developments over the last few decades that resulted collectively in bringing climate issues to the forefront (Zillman 2009). The significant milestones were the following:

- The International Geophysical Year (IGY) of 1957-58, which was a major effort towards international cooperation in science, during which regular observations of many new geophysical parameters were started. A significant outcome of the IGY was the setting up of an observatory at Mauna Loa in Hawaii for the continuous monitoring of atmospheric CO_2. This station is still functioning and providing data that is fundamental to our understanding of global warming.

- The launch of the first weather satellite in 1960 by the U. S. which was the precursor to many more meteorological satellites of increasing complexity and capability being launched by the U. S. and other countries including India. The current constellation of satellites engaged in remote sensing of land, ocean and atmosphere is giving us information about climate change that it would have been otherwise impossible to obtain.

- The U. N. General Assembly Resolution of 1961 that called upon the nations of the world to collaborate through the World Meteorological Organization (WMO) and the International Council for Science (ICSU) for monitoring and predicting the weather. This spawned the birth of the World Weather Watch (WWW) and the Global Atmospheric Research Programme (GARP), followed later by the World Climate Programme (WCP) and the Global Climate Observing System (GCOS).

- The establishment of the Intergovernmental Panel on Climate Change (IPCC) in 1988, jointly by the WMO and United Nations Environment Programme (UNEP), to serve as an authoritative source of scientific information on climate change and its impacts and to bring national governments into the process of reviewing such information.

- The Earth Summit at Rio de Janeiro in 1992 at which the nations of the world agreed to the establishment of the U. N. Framework Convention on Climate Change (UNFCCC), which came into being in 1994. The parties to the UNFCCC made specific commitments towards reduction of greenhouse gases in specified time frames.

- The Kyoto Protocol of 1997 under which legally binding commitments were stipulated for the developed countries and the Clean Development Mechanism (CDM) was set up for the benefit of the developing countries.

1.3 Energy Balance of the Earth

The sun is the source of all the energy that is received by planet earth. This energy drives the earth's atmosphere and oceans and sustains all life on earth. Simple physical reasoning tells us that the energy received by the earth must eventually be sent back by the earth to space, because an imbalance between the two will result in an increase or decrease in the earth's temperature. For the earth's temperature to remain stable, there has to be a radiative and energy balance within the earth-ocean-atmosphere system. This balance is required to be maintained only in the long term and for the earth as a whole, as on the smaller time and space scales, imbalances abound. In fact it is these sources and sinks of radiation and energy that are responsible for the movements of the atmosphere and ocean All this was known hundreds of years ago and the individual components of the earth's radiation and energy budgets had been quantified quite precisely. Even today the satellites that we have are not capable of measuring most of the energy budget components except the incoming and outgoing fluxes at the top of the atmosphere.

At any level in the atmosphere above a given place and at a given time there are some fluxes coming downward towards the earth's surface and some fluxes going upwards away from the surface towards space. Thus in a finite layer of the atmosphere between two horizontal levels, there are upward and downward fluxes at both its boundaries. If the resultant or net flux is positive, the layer will experience heating and if the net flux is negative there will be cooling. However, for the earth-atmosphere-ocean system as whole, all the individual component fluxes must, on the climate scale, get neutralised and the net flux must be zero. If on the climate scale, the net flux has a positive residual, it would amount to retention of heat by the climate system and an increase in the average temperature. If on the climate scale, the system loses more heat than in receives, that would result in cooling and a decrease in the average temperature. So if the overall radiative energy balance of the earth's climate system is disturbed, the result would be global warming or global cooling depending upon the nature of what is called the radiative forcing. The source of the radiative forcing could be external to the climate system, or it could lie within it. The internal forcing could again be is it natual or anthropogenic.

Therefore, in order to examine the nature, magnitude and cause of the current global warming, it is necessary in the first place to go into some details about how the earth's energy balance is maintained. The solar energy that enters the earth's atmosphere undergoes a series of transformations as it traverses through the atmosphere towards the land and ocean surface. It is attenuated because of reflection at the cloud tops, scattering by aerosols and absorption by some of the minor constituent gases like ozone. Some of the radiation that manages to reach the surface is again partially reflected and the rest gets absorbed by the ground or ocean. The solar energy absorbed by the earth, atmosphere and ocean gets converted to other forms of energy like sensible heat, latent heat, kinetic energy and potential energy. The net balance of the various energy and radiation balance componants determines the temperature at the earth's surface, which then emits longwave radiation as per its equlibrium temperature..

This terrestrial radiation travels upwards and gets absorbed by clouds and some other minor constituent gases like water vapour, CO_2, methane (CH_4) and ozone (O_3). The clouds and gases that absorb it also emit longwave radiation at their own temperatures, in both upward and downward directions.

The major factors that control or alter the energy budget are therefore:

- albedo or reflectance properties of the earth's land surface, ocean and cloud tops

- distribution of clouds of various types and at various heights

- vertical profiles of the concentration of gases which absorb longwave radiation like CO_2, water vapour, ozone and others

- distrubution of aerosols, dust and particulate matter in the atmosphere, which scatter solar radiation and may also absorb longwave radiation

- vertical temperature profile of the atmosphere

If the composition and properties of the atmosphere are known, the contribution of the above factors can be estimated by applying the principles and laws of radiative transfer. Figure 1.3.1 shows the magnitudes of the earth's energy budget components when averaged globally over the year as a whole. In reality of course, the composition and properties of the atmosphere are not fully known. Many of the parameters listed above have not been measured over the ocean, or not with the desired resolution and accuracy. The radiative properties of different types of aerosols are not yet completely known. Interactions of clouds and aerosols, and between clouds and the atmosphere, are also to be fully understood. The absolute magnitudes of the different fluxes (indicated in Watts per sq metre (or Wm^{-2}) shown in Figure 1.3.1 are therefore subject to many uncertainties. Nevertheless, a comparison of their relative magnitudes helps in understanding the potential importance of various radiative forcings in altering the future climate of the earth. In IPCC parlance, radiative forcing values are for 2005 relative to pre-industrial conditions defined at 1750 and are expressed in Wm^{-2}.

The radiation received from the sun outside the earth's atmosphere is called the solar constant. It was presumed to be a constant but it is now known to have slight annual and interannual variations. It has an average value of 1366 $W\ m^{-2}$ when measured perpendicular to the solar beam at the mean sun-earth distance. However, at any time, half the earth is dark and the sun's rays are inclined to the earth's surface at different angles depending upon latitude and season. So the global annual average incoming solar radiation is 342 Wm^{-2} as shown in Figure 1.3.1. It is more meaningful and easier, however, to look at the various values in this figure in terms of percentages rather than in absolute magnitudes in Wm^{-2}.

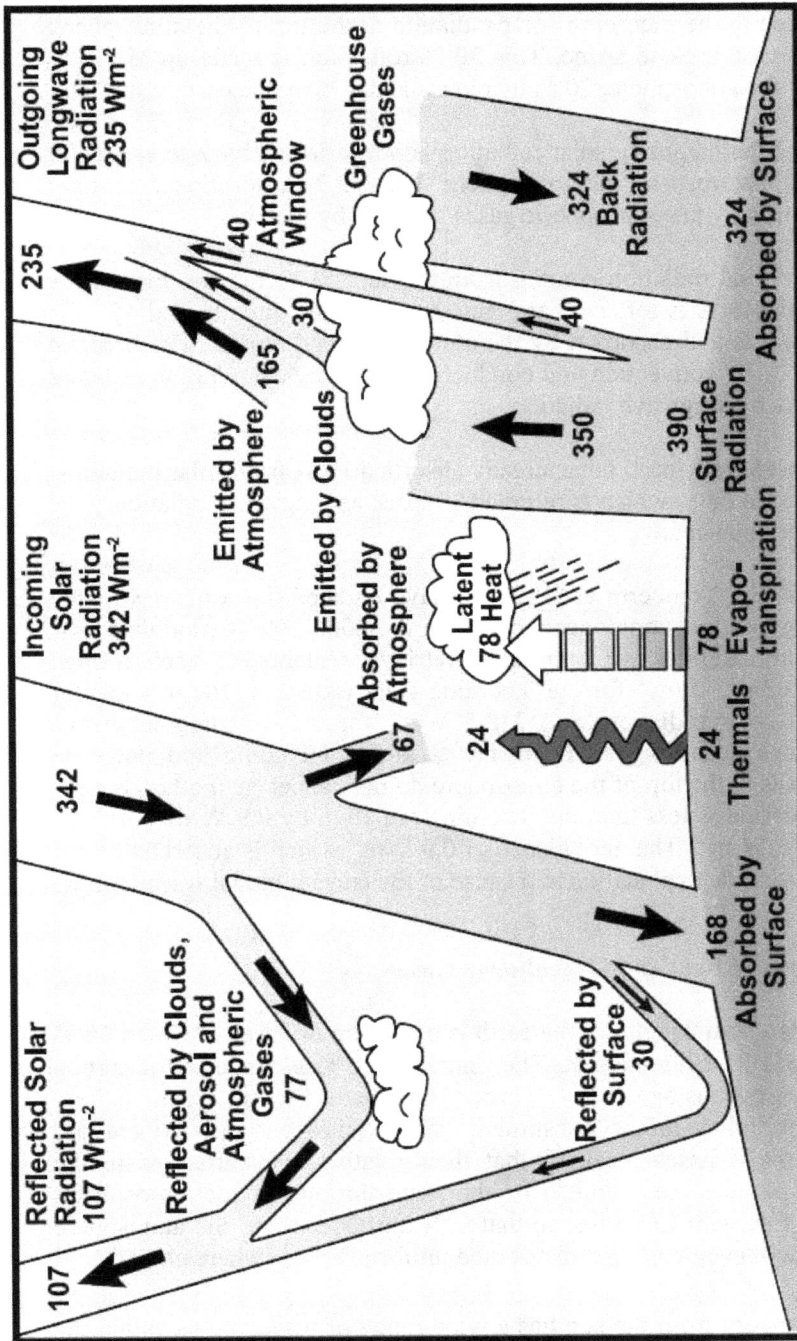

Figure 1.3.1 Earth's energy budget (Source: IPCC AR4, Le Treut et al 2007)

With reference to the incoming solar radiation at the top of the atmosphere, 30 % is reflected back to space. This 30 % reflection is made up of a 6 % reflection by the atmosphere, 20 % by clouds and 4 % by the earth's surface.

Thus 70 % of the incoming solar radiation is not reflected back to space. Of this, 19 % is absorbed by the atmosphere. This 19 % absorption comprises 16 % absorption by the atmospheric gases and 3 % by clouds.

So out of the total radiation coming from the sun, 51 % reaches the earth's surface. Of this 6 % is reflected and gets lost to space directly and 45 % is returned upwards and absorbed by the atmosphere and clouds. The breakup of 45 % is 7 % by convection and conduction, 23 % by evaporation as latent heat and 15 % by longwave radiation.

The atmosphere and clouds have already absorbed 19 % from solar radiation, making a total of 64 % which is returned to space as longwave radiation. The budget is thus balanced.

In a recent study, Trenberth et al (2009) have updated the estimates of the various energy budget components based on the 2000-2004 period data now available from satellites with improved retrieval techniques. Their refined estimates are 341.3 Wm^{-2} for the incoming solar radiation, 101.9 Wm^{-2} for the reflected solar radiation, and 238.5 Wm^{-2} for the outgoing longwave radiation. Thus according to Trenberth et al the incoming and outgoing radiative fluxes at the top of the atmosphere do not cancel each other, but the outgoing radiation is less than the incoming radiation by 0.9 W m^{-2}, with an error of ±0.15 W m^{-2}. The net balance of 0.9 Wm^{-2} which is absorbed by the surface is a possible explanation and cause of the current global warming.

1.4 Greenhouse Effect and Greenhouse Gases

The atmosphere that envelopes the earth is a mixture of gases of which 78 % is nitrogen and 21 % is oxygen. The remaining 1 % is made up of carbon dioxide (CO_2), nitrous oxide (N_2O), ozone (O_3), water vapour, argon, helium, hydrogen and other minor constituent gases. Nitrogen and oxygen are thoroughly mixed gases meaning that their relative concentrations in the atmospheric mixture are found to be the same throughout the atmosphere. The same is true with CO_2 also, so that CO_2 introduced into the atmosphere at any place will eventually get distributed uniformly everywhere else.

The radiant energy from the sun has a wide range of wavelengths called the electromagnetic spectrum (Table 1.4.1). As this energy passes through the earth's atmosphere towards the surface, it interacts with the molecules of the atmospheric gases differently at different wavelengths. Ozone, which resides in the atmosphere at a height of about 25 km, is the first to interact and it absorbs the sun's ultraviolet radiation. The ozone layer primarily acts as shield that protects living beings on earth from the deadly effects of ultraviolet radiation like skin cancer and other diseases.

Except ozone, the other atmospheric gases do not absorb solar radiation directly. Solar radiation in the visible wavelengths is reflected at cloud tops and at the earth's surface and it is scattered by aerosols and particulate matter in the atmosphere. The portion that reaches the surface is absorbed and reradiated upwards into the atmosphere. It is this terrestrial or longwave radiation that gets absorbed by the atmospheric gases, particularly water vapour, CO_2, methane and N_2O.

Table 1.4.1 Electromagnetic Spectrum

Wavelength		Wavelength	
10^{-6} nm		1 mm	Millimetre Waves (mm)
10^{-5} nm	Gamma Rays (MeV)	1 cm	Microwaves (cm, GHz)
10^{-4} nm		10 cm	
10^{-3} nm		1 m	
10^{-2} nm		10 m	
10^{-1} nm		100 m	
1 nm	X-Rays (Å)	1 km	
10 nm		10 km	Radio Waves (MHz, kHz)
100 nm	Ultra-Violet (nm),	100 km	
1 μ	Visible, Near Infra-Red (μ)	10^3 km	
10 μ	Thermal Infra-Red (μ)	10^4 km	
100 μ	Far Infra-Red (μ)	10^5 km	

The heat absorbing action of these gases is akin to what happens in a greenhouse commonly used for growing plants. The glass cover of the greenhouse is transparent to solar radiation and so it can enter the greenhouse. The surface radiates longwave radiation which, however, the glass does not allow to leave. The heat energy thus gets trapped within the greenhouse and the temperature rises. Because of the greenhouse analogy, the heat-absorbing gases in the atmosphere such as water vapour, CO_2, N_2O and methane have come to be known as GreenHouse Gases or GHGs for short. There are other minor GHGs in the atmosphere as well, like hydro-chloro-fluoro-carbons (HCFCs), but their concentrations are much smaller.

If the climate system is considered as a whole, there is a considerable exchange of GHGs within the system, like between the ocean and the atmosphere, and between plants and the atmosphere. The process of photosynthesis in plants results in a constant removal of CO_2 from the atmosphere and converting it into carbohydrates, while in the process of respiration in humans and other living beings, CO_2 is being exhaled into the atmosphere. The amount of water vapour in the atmosphere or moisture is highly variable, and the processes of evaporation and condensation are constantly going on.

If GHGs are added into the atmosphere by anthropogenic processes, particularly the release of CO_2 through burning of fossil fuels, the natural greenhouse effect of the atmosphere will get more pronounced and result in increased temperatures. The climate system also has its own feedback mechanisms, both positive and negative. For example, as the atmosphere gets warmer its moisture content increases, and this can work towards a further greenhouse warming.

Cloud-atmosphere and cloud-aerosol interactions are also important in this context. Besides GHGs, clouds are also very good absorbers of longwave radiation and have their own greenhouse effect. Thus an increase in water vapour because of GHG warming, would be conducive to the process of cloud formation and so lead to further warming. On the other hand, clouds have a high albedo, and a lot of solar radiation gets reflected at cloud tops, so increase in cloudiness could result in more solar energy getting reflected back to space and cause a cooling. These mechanisms are complex and need to be further understood.

In 2008, the global mean concentrations of the three most important greenhouse gases, CO_2, CH_4 and N_2O, stood at 385.2 ppm, 1797 ppb and 321.8 ppb respectively as per the latest data made available by the WMO (Table 1.4.2). The unit ppm/ppb stands for the number of molecules of the GHG per million/billion molecules of dry air. Prior to 1750 or the beginning of the era of the industrial revolution, these GHGs had nearly stable concentrations of 280 ppm, 700 ppb and 270 ppb respectively, for thousands of years (Figure 1.4.1). The increase in the concentrations from 1750 to the present is thus of the order of 38% for CO_2, 157% for methane and 19% for N_2O.

The annual global mean concentration of CO_2 represents a balance among various components of the carbon cycle. Huge seasonal exchanges of carbon and carbon dioxide take place between the atmosphere and the ocean on one hand, and between the atmosphere and biosphere through photosynthesis and respiration on the other. The 38% increase in CO_2 concentration is attributable primarily to the emissions from combustion of fossil fuels and to a lesser extent to deforestation, although some of the increase has been offset by the removal of CO_2 by the biosphere and oceans. CO_2 is the single most important infrared radiation absorbing, anthropogenic gas in the atmosphere and is responsible for 63 % of the total radiative forcing of the earth by long-lived greenhouse gases.

Table 1.4.2 Trends in global mean concentrations of the three major greenhouse gases (Source: WMO 2009a)

	Carbon Dioxide (CO2)	Methane (CH4)	Nitrous Oxide (N2O)
Mean global concentration in 2008	385.2 ppm	1797 ppb	321.8 ppb
Mean global concentration in 1750	280 ppm	700 ppb	270 ppb
Increase in mean global concentration from 1750 to 2008	38%	157%	19%
Mean annual absolute increase from 1999 to 2008	1.93 ppm	2.5 ppb	0.78 ppb

The global mean concentration of methane is determined by a variety of terrestrial and atmospheric processes. About 40 % of atmospheric methane originates naturally from wetlands and termites and the remaining 60% comes from anthropogenic sources like burning of fossil fuels, rice cultivation, ruminant animals, or landfills. Methane has a complex cycle and it gets removed from the atmosphere primarily by reaction with the hydroxyl radical (OH). Its residence time in the atmosphere is of the order of 9 years. The 157% increase in atmospheric methane since 1750 to the present is attributable mainly to increasing emissions from anthropogenic sources. Methane contributes 18.5 % of the direct radiative forcing due to long-lived greenhouse gases affected by human activities. It also interacts with tropospheric ozone and stratospheric water vapour and can therefore influence climate indirectly.

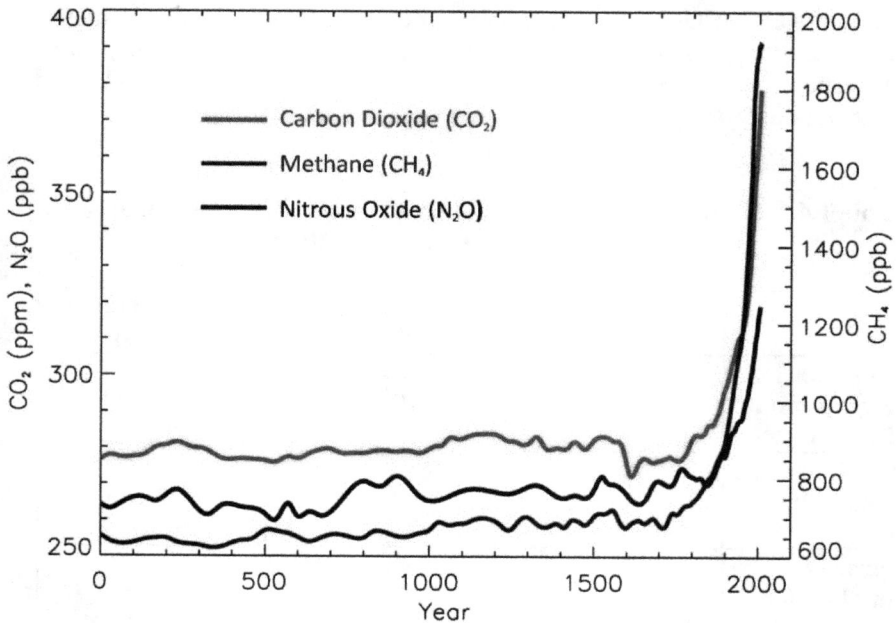

Figure 1.4.1 Concentrations of GHGs over the years 0 to 2005
(Source: IPCC AR4, Solomon et al 2007)

In terms of the earth's radiative forcing, the third important GHG is Nitrous oxide (N_2O) which contributes 6.2 % of the total. The increase in the global mean concentration of N_2O since the beginning of the industrial era has been 19 %. N_2O is emitted into the atmosphere from natural as well as anthropogenic sources over land and ocean, combustion of fossil fuels, biomass burning, use of fertilizers in agriculture, and different industrial processes. One-third of its total emission is from anthropogenic sources. It is removed from the atmosphere by photochemical processes in the stratosphere.

Figure 1.4.2 shows global averages of radiative forcings of climate as of 2005 for anthropogenic CO_2, methane, N_2O and other important agents and mechanisms. According to the IPCC Fourth Assessment Report (Solomon et al 2007), or AR4 for short, the understanding of anthropogenic warming and cooling influences on climate has improved since its third assessment which was made in 2001. It can be said with great confidence that globally averaged net effect of human activities since 1750 has been one of warming, with a radiative forcing of +1.6 [+0.6 to +2.4] Wm^{-2}.

Radiative Forcing Components

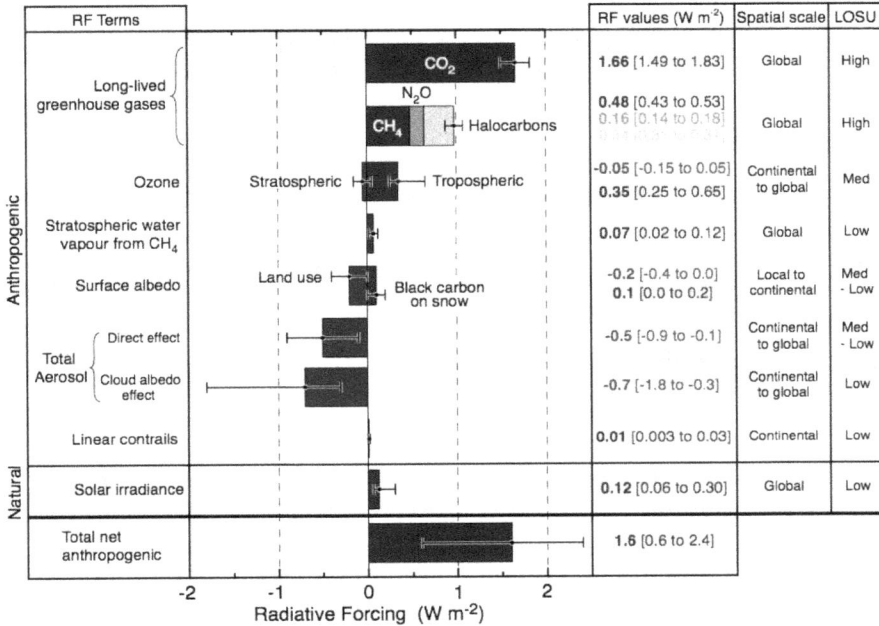

RF Terms			RF values (W m^{-2})	Spatial scale	LOSU
Long-lived greenhouse gases	CO$_2$		1.66 [1.49 to 1.83]	Global	High
	N$_2$O		0.48 [0.43 to 0.53]		
	CH$_4$	Halocarbons	0.16 [0.14 to 0.18]	Global	High
Ozone	Stratospheric	Tropospheric	-0.05 [-0.15 to 0.05] 0.35 [0.25 to 0.65]	Continental to global	Med
Stratospheric water vapour from CH$_4$			0.07 [0.02 to 0.12]	Global	Low
Surface albedo	Land use	Black carbon on snow	-0.2 [-0.4 to 0.0] 0.1 [0.0 to 0.2]	Local to continental	Med - Low
Total Aerosol — Direct effect			-0.5 [-0.9 to -0.1]	Continental to global	Med - Low
Cloud albedo effect			-0.7 [-1.8 to -0.3]	Continental to global	Low
Linear contrails			0.01 [0.003 to 0.03]	Continental	Low
Solar irradiance			0.12 [0.06 to 0.30]	Global	Low
Total net anthropogenic			1.6 [0.6 to 2.4]		

Radiative Forcing (W m^{-2})

Figure 1.4.2 Global average radiative forcings of climate in 2005. Values and error bars are indicated in Wm^{-2}. LOSU is the assessed level of scientific understanding. (Source: IPCC AR4, Solomon et al 2007)

The combined radiative forcing due to increases in CO$_2$, methane and N$_2$O is +2.30 [+2.07 to +2.53] Wm^{-2}, and its rate of increase during the industrial era is very likely to have been unprecedented in more than 10,000 years. The carbon dioxide radiative forcing increased by 20 % from 1995 to 2005, the largest change for any decade in at least the last 200 years.

Anthropogenic contributions to aerosols, primarily sulphate, organic carbon, black carbon, nitrate and dust, together produce a cooling effect, with a total direct radiative forcing of -0.5 [-0.9 to -0.1] Wm^{-2} and an indirect cloud albedo forcing of -0.7 [-1.8 to -0.3] Wm^{-2}. These forcings are now better understood than at the time of the third IPCC assessment thanks to improved measurements and more comprehensive modelling, but these still remain the dominant uncertainty in radiative forcing. Aerosols also influence cloud lifetime and precipitation. Significant anthropogenic contributions to radiative forcing come from several other sources. Tropospheric ozone changes due to emissions of ozone-forming chemicals (nitrogen oxides, carbon monoxide, and hydrocarbons) contribute +0.35 [+0.25 to +0.65] Wm^{-2}. The direct radiative forcing due to changes in halocarbons is +0.34 [+0.31 to +0.37] Wm^{-2}. Changes in surface albedo, due to land cover changes and deposition of black carbon aerosols on snow, exert respective forcings of -0.2 [-0.4 to 0.0] and +0.1 [0.0 to +0.2] Wm^{-2}. Changes in solar irradiance since 1750 are estimated to cause a radiative forcing of +0.12 [+0.06 to +0.30] Wm^{-2}.

1.5 Observed Global Warming and Climate Change

One of the earliest indications of global warming were reported by Hansen et al (1981) who found that the average global temperature had increased by 0.2 °C from the mid-1960s to 1980, yielding a warming of 0.4 °C in the previous 100 years. They argued that this temperature rise was consistent with calculations based upon the observed increase of atmospheric carbon dioxide. They also suggested that possible changes in solar luminosity and aerosols thrown into the atmosphere by volcanic activity may be the causes of observed fluctuations about the mean trend of increasing temperature. They predicted that global warming would continue in the 1980s and that anthropogenic carbon dioxide warming would exceed natural climate variability by the end of the 20th century.

The compilation of credible historical global temperature data became crucial for carrying on further and more definitive investigations of global warming. This need was addressed by three different institutions, the U. K. Met Office Hadley Centre, NOAA National Climatic Data Center and NASA Goddard Institute of Space Studies, who are now working in parallel to compile, maintain and constantly update the global surface temperature data. The three centres, however, use different averaging periods for deriving the mean global temperature and different methodologies for computing the annual global average. Thus the three data series match in a qualitative sense, but there are differences in the numerical values of the global warming parameters, ranking of warm years and other statistics.

IPCC's Third Assessment Report published in 2001 had stated that the average surface temperature of the earth had increased by 0.6 ± 0.2 °C over the twentieth century (1901-2000). The Fourth Assessment Report of the IPCC (Solomon et al 2007) released in 2007, reconfirmed unequivocally that global warming was continuing and at an even faster rate than before. Over the 100-year period 1906-2005 the temperature had risen by 0.74 ± 0.18 °C which was higher than that for the 100-year period 1901-2000. The total temperature increase from 1850-1899 to 2001-2005 was 0.76 ± 0.19°C. The years 2005 and 1998 were the two warmest years in the instrumental global surface air temperature record since 1850. Surface temperatures in 1998 were enhanced by the major 1997-1998 El Nino but no such strong anomaly was present in 2005. Eleven of the 12 years 1995-2006 with the exception of 1996, ranked among the 12 warmest years on record since 1850 up to 2006 (Figure 1.5.1).

It is important to give a careful attention to the magnitude of global warming. If over the 100-year period 1906-2005, the temperature rise has been 0.74 with an uncertainty of $\pm$ 0.18 °C, it means that that the temperature rise could have been anywhere between 0.56 and 0.92 °C over 100 years, or between 0.056 and 0.092 °C in a year on an average. So we are dealing here with extremely small quantities with relatively large error bars.

Annual temperature anomalies are derived as departures of the annual mean temperature from the normal or long-term average temperature. There are serious problems in computing both these parameters. The earth's long-term average surface temperature which is used as the basis for computing annual temperature anomalies is not uniquely defined. It is approximately 14 °C but not exactly, as the average value changes when different averaging periods are used. The annual global average temperature is derived from data recorded through a host of observing systems, like surface observatories, ships, ocean buoys, and satellite retrievals. Each of these have different errors and uncertainties associated with them and are not always compatible. Obtaining one single value of temperature averaged over the entire earth and over a year, is not a straightforward task, nor is there a unique internationally accepted practice for doing so. This issue is discussed in detail later in this book in Section 2.3.

While Figure 1.5.1 shows the global warming trend for the globe as a whole, Figure 1.5.2 shows the spatial variation across the globe of the annual surface temperature changes between 1901 and 2005. The warming trend is statistically significant over most regions with the exception of an area south of Greenland and southeastern U. S.

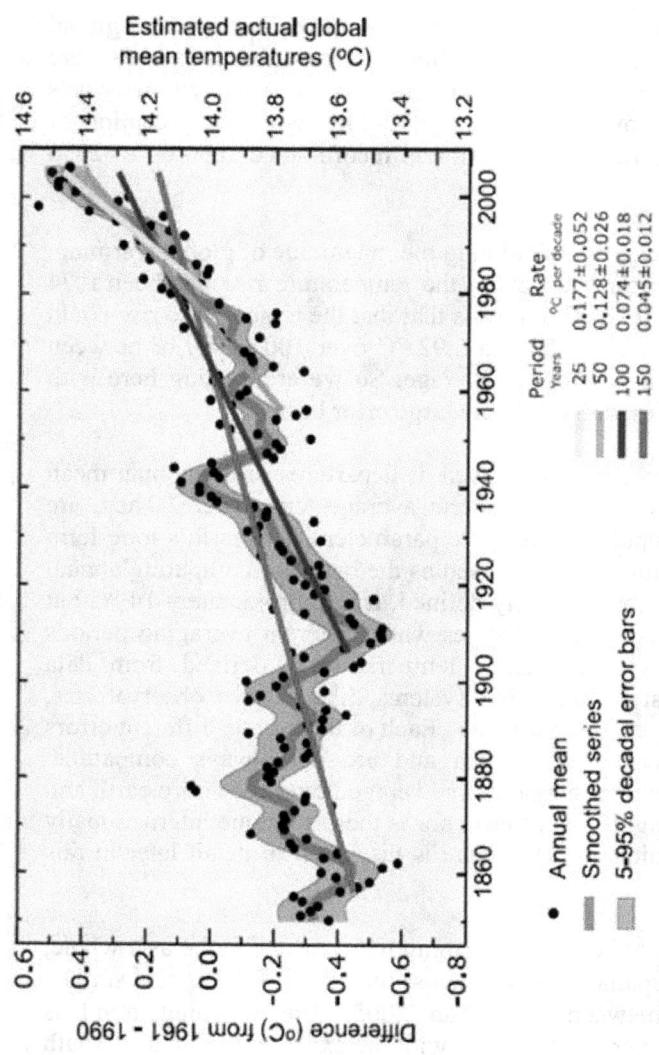

Figure 1.5.1 Annual global mean temperatures (black dots). The left hand axis shows temperature anomalies relative to the 1961-1990 average and the right hand axis shows estimated actual temperatures, both in °C. Coloured lines show linear trends over various periods. The smooth blue curve shows decadal variations with the decadal 90% error range shown as a pale blue band about that line. (Source: IPCC AR4, Solomon et al 2007)

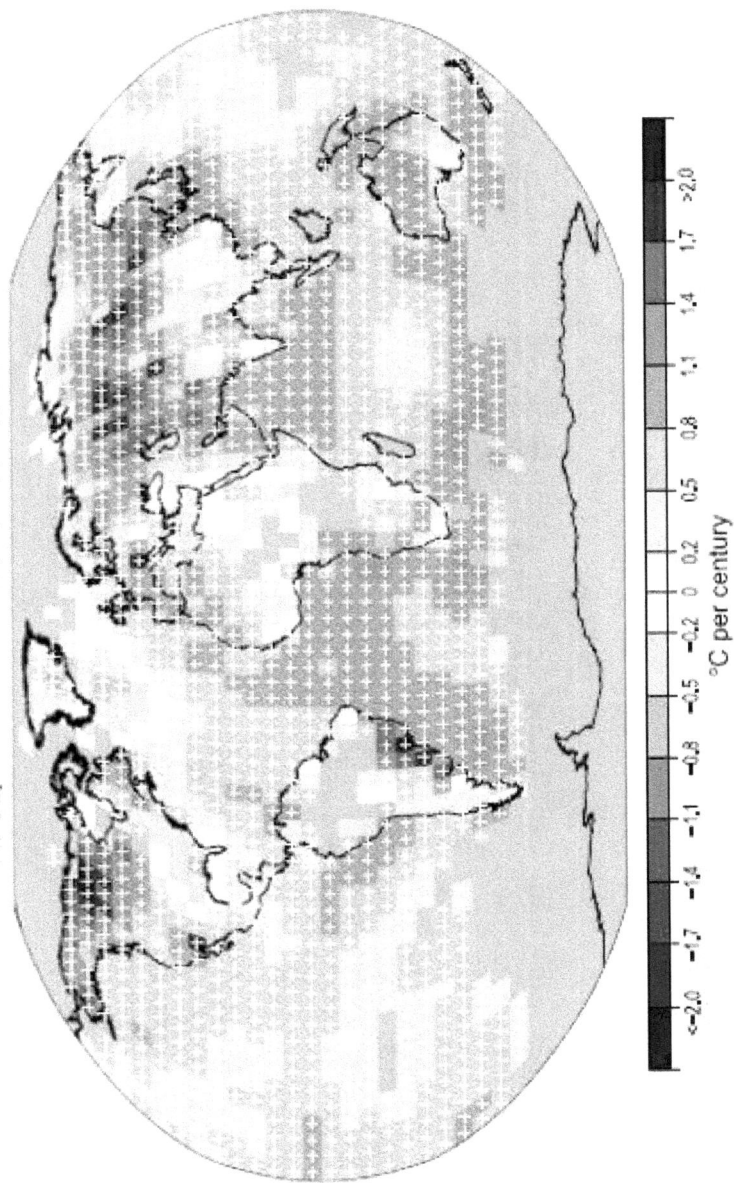

Figure 1.5.2 Linear trend of annual temperatures over the period 1901-2005 in °C per century Areas in grey have insufficient data. Trends significant at the 5% level are indicated by white + marks. (Source: IPCC AR4 Solomon et al 2007)

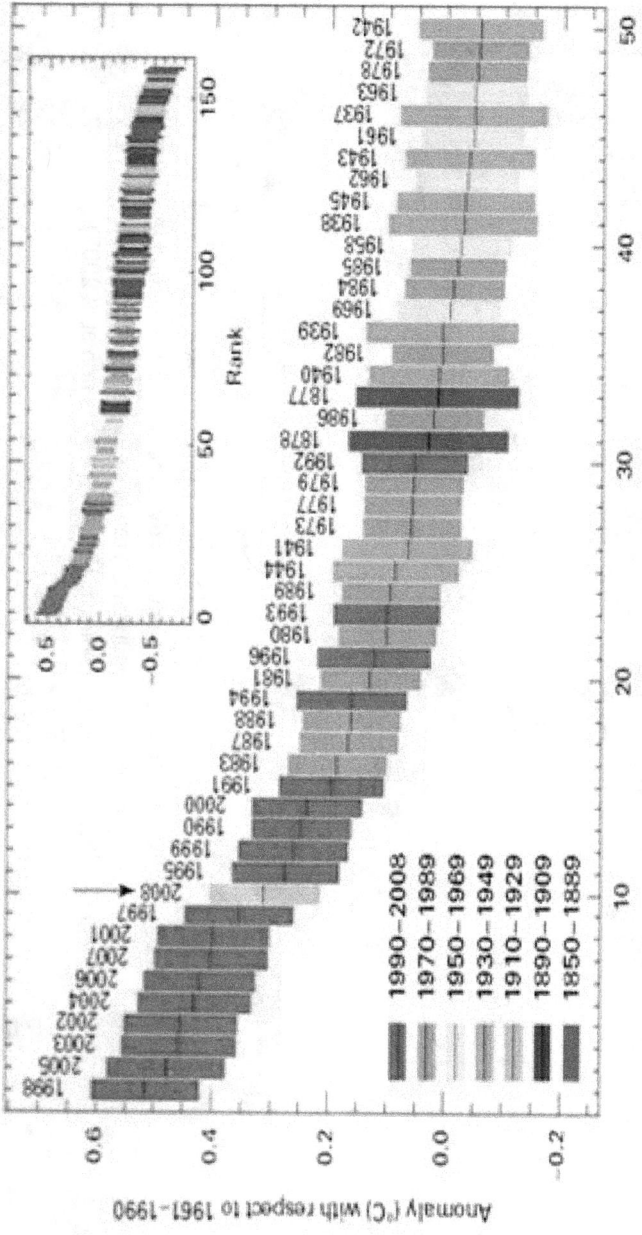

Figure 1.5.3 Annual global combined land surface air temperature and sea surface temperature anomalies (°C) relative to the average for 1961-1990, ranked for the warmest 50 years since 1850. Inset shows ranking for all years from 1850 onwards. The size of the bars indicates the 95 per cent confident limits associated with each year. (Source: WMO web site www.wmo.int)

The warming is strongest over the interior parts of the Asian continent, and northwestern parts of north America and over some mid-latitude ocean regions of the southern hemisphere and southeastern Brazil. It is interesting to see that there is considerable variation of the warming over the Indian subcontinent from near zero over southern India to 0.5-1.0 °C per century over central and eastern India.

No single location follows the global average, and the only way to monitor the globe with any confidence is to include observations from as many diverse places as possible (Solomon et al 2007). The error bars assume even further importance with regard to the spatial mapping of warming trend.

Figure 1.5.3 shows the annual global combined land surface air temperature and sea surface temperature anomalies (°C) relative to the average for 1961-1990, ranked for the warmest 50 years since 1850. The inset shows the ranking for all years from 1850 onwards. The size of the bars indicates the 95 per cent confidence limits associated with each year.

The Fourth Assessment Report of the IPCC says categorically that warming of the climate system is unequivocal, as is now evident from observations of increase in global average air and ocean temperatures, widespread melting of snow and ice, and rising global average sea level (Solomon et al 2007). Besides the basic evidence of global warming in the surface temperatures, IPCC AR4 cites several indications of change as seen in a wide variety of climate parameters:

- New analyses of balloon-borne and satellite measurements of lower and middle tropospheric temperatures show warming rates that are similar to those of the surface temperature record and are consistent within their respective uncertainties.

- The average atmospheric water vapour content has increased since at least the 1980s over land and ocean as well as in the upper troposphere. The increase is broadly consistent with the extra water vapour that warmer air can hold.

- Observations since 1961 show that the average temperature of the global ocean has increased to depths of at least 3 km and that the ocean has been absorbing more than 80 % of the heat added to the climate system. Such warming causes sea water to expand, contributing to sea level rise. Widespread decreases in glaciers and ice caps have contributed to sea level rise. Global average sea level rose at an average rate of 1.8 mm/yr over 1961-2003 but the rate was faster over 1993-2003, about 3.1 mm/yr.

- On continental, regional, and ocean basin scales, numerous long-term changes in climate have been observed. These include changes in Arctic temperatures and ice, widespread changes in precipitation amounts, ocean salinity, wind patterns and aspects of extreme weather including droughts, heavy precipitation, heat waves and the intensity of tropical cyclones. Satellite data since 1978 show that annual average Arctic sea ice extent has shrunk by 2.7 % per decade.

- Long-term trends from 1900 to 2005 have been observed in precipitation amount over many large regions. Significantly increased precipitation has been observed in eastern parts of North and South America, northern Europe and northern and central Asia. Drying has been observed in the Sahel, the Mediterranean, southern Africa and parts of southern Asia. Precipitation is highly variable spatially and temporally, and data are limited in some regions. Long-term trends have not been observed for the other large regions assessed.

- The frequency of heavy precipitation events has increased over most land areas, consistent with warming and observed increases of atmospheric water vapour.

- Widespread changes in extreme temperatures have been observed over the last 50 years. Cold days, cold nights and frost have become less frequent, while hot days, hot nights, and heat waves have become more frequent.

- There is observational evidence for an increase of intense tropical cyclone activity in the North Atlantic since about 1970, correlated with increases of tropical sea surface temperatures.

- Mid-latitude westerly winds have strengthened in both hemispheres since the 1960s.

1.6 Other Causes of Change in the Earth's Temperature

There is geological and paleoclimatological evidence that climate changes have occurred on the earth in the past on various time scales even when anthropogenic activities could not possibly have been of any consequence. The climate changes in the past could have also occurred due to a disruption of the climate equilibrium but for different reasons.

It is known that ice ages occurred at regular intervals in the past three million years, interspersed by warm periods. There is strong evidence that these are linked to regular variations in the earth's orbit around the sun, which are called the Milankovitch cycles. These cycles alter the latitudinal and seasonal variation of the incoming solar radiation at the top of the atmosphere, but not the global annual mean, and they can be calculated with astronomical precision. The exact linkage of the Milankovich cycles with the onset and end of the ice ages is however not completely understood.

Carbon dioxide (CO_2) is also believed to have played an important role in the onset and cessation of the ice ages as the Antarctic ice core data show a good correlation of the CO_2 concentrations with the temperature anomalies (Section 2.8).

It is also possible that the incoming solar radiation could have changed because of a change in the sun's emission itself. Solar sunspot activity has a high degree of variability, the most common being the regular 11-year cycle. Recent measurements have shown that the sunspot cycles are associated with small changes in the solar output also. Long-term changes on the solar output cannot therefore be ruled out as a cause of climate change. Volcanic activity could be another cause of climate change as volcanic outpourings can form a layer in the earth's atmosphere blocking the solar radiation and creating a cooling effect.

Another hypothesis (Carslaw et al 2002) is that cosmic rays could also exert an indirect influence on the earth's climate through a change in the cloudiness. This stems from an observed correlation between cosmic ray intensity and the earth's average cloud cover over the course of one solar cycle. The physical connection between cosmic rays and clouds is not yet proven and needs to be further investigated. Pierce et al (2009) have tried to explain that it is theoretically possible for cosmic rays to generate more new particles and higher concentrations of cloud condensation nuclei (CCN) and therefore be conducive to the formation of clouds. However their model simulations have shown that the changes in the CCN on account of cosmic rays during one or two solar cycles are two orders of magnitude lower than required to explain the observed changes in cloud properties and too small to play a significant role in the current climate change.

1.7 Antarctic Ozone Hole

The recent depletion of atmospheric ozone, first noticed in the early 1980s, and particularly the discovery of the annually recurring phenomenon called the Antarctic ozone hole, has become a matter of global concern. Ozone is a very minor constituent of the earth's atmosphere and about 90% of it resides in the stratosphere at an altitude of about 25 km. The ozone layer however plays a significant role in protecting living beings on earth from the harmful effects of the ultraviolet radiation that comes from the sun. Any depletion of stratospheric ozone by possible human-induced processes is going to make us increasingly vulnerable to the solar ultraviolet radiation and prone to its adverse health effects.

The other property of ozone is that it acts as a greenhouse gas in the troposphere, though its radiative forcing is very small compared to that of CO_2 which is the prime greenhouse gas. An increase in the tropospheric ozone concentration resulting from anthropogenic air pollution will therefore contribute to a warming of the atmosphere. For both these reasons, ozone has assumed importance in the current context of global warming and climate change.

The earliest satellite measurements of atmospheric ozone began in 1978 with the Total Ozone Mapping Spectrometer (TOMS) sent on the Nimbus-7 satellite. The instrument has since been flown on different satellites of Russia, U. S. and Japan and a long data series has been generated. TOMS measures the total ozone in the atmospheric column by observing both incoming solar energy and backscattered ultraviolet (UV) radiation at six different wavelengths. Backscattered radiation is solar radiation that has penetrated into the earth's atmosphere and has then been scattered by air molecules and clouds back through the atmosphere to the satellite sensors. Along that path, a fraction of the UV radiation is absorbed by ozone. By comparing the amount of backscattered radiation to observations of incoming solar energy at identical wavelengths, the amount of ozone in the atmospheric column can be indirectly estimated. It is also possible to calculate the earth's UV albedo, the ratio of light reflected by the earth to what it receives. Changes in albedo at the selected wavelengths can be used to derive the amount of atmospheric ozone.

Currently, ozone is being regularly monitored globally by several space-borne instruments such as the Ozone Monitoring Instrument (OMI) on-board NASA's Aura satellite, the Global Ozone Monitoring Experiment (GOME) on the European ERS-2 spacecraft, and the Solar Backscatter Ultraviolet (SBU) instrument on the NOAA-16 satellite.

The first significant depletions in ozone concentrations over the Antarctic region in the months of September to November were noticed in the early 1980s through ground-based observations. Soon afterwards, this was confirmed by satellite measurements and the area of ozone depletion was seen to extend over a large region around the South Pole. This area came to be known as the Antarctic ozone hole.

Several ground stations have been established in the Antarctic region for making regular measurements of surface and column ozone and also obtain vertical profiles of ozone concentrations through balloon-based ozonesondes. This combination of surface, balloon and satellite data provides a complete picture of the ozone hole and helps the scientific community to keep a close watch on the interannual variation of its properties (Figure 1.7.1).

The Antarctic ozone hole broke all records for both area and depth in September 2006. Figure 1.7.2 is the OMI image of 24 September 2006. The 2006 ozone hole also persisted longer than in any previous year. Figure 1.7.3 is the OMI image of 10 September 2009 which shows that the ozone hole of 2009 is comparable to ozone depletions over the past decade. Ozone concentrations are indicated in Dobson Units, with purple and blue areas depicting severe deficits of ozone as per the colour scale in Figure 1.7.2.

That the ozone hole forms only over the Antarctic region and not anywhere else, is attributable to the peculiar meteorological conditions which prevail there. Compared to the troposphere, water vapour concentrations in the stratospheric layers of the atmosphere are extremely low and cloud formation is normally not possible. However, over the Antarctic region, when stratospheric temperatures drop to below -78 °C, clouds can form. These are called polar stratospheric clouds and they consist of a mixture of water and nitric acid. In this process, chemical reactions occur that transform harmless halogen compounds like HCl or HB into active chlorine and bromine species such as ClO and BrO. These active forms of chlorine and bromine cause rapid ozone depletion in sunlit conditions. When temperatures drop below -85 °C, ice clouds can form, in which case the ozone depletion is more severe. The other important factor is the polar vortex which is a large low pressure system in which strong stratospheric winds of the polar jet circle the Antarctic continent. The region poleward of the polar jet has the lowest temperatures. The unique combination of the chemical processes of cloud formation and the polar vortex, which exists nowhere else in the world, results in the Antarctic ozone hole. The ozone hole begins to grow in August and reaches its largest area in late September to early October after which it shrinks again.

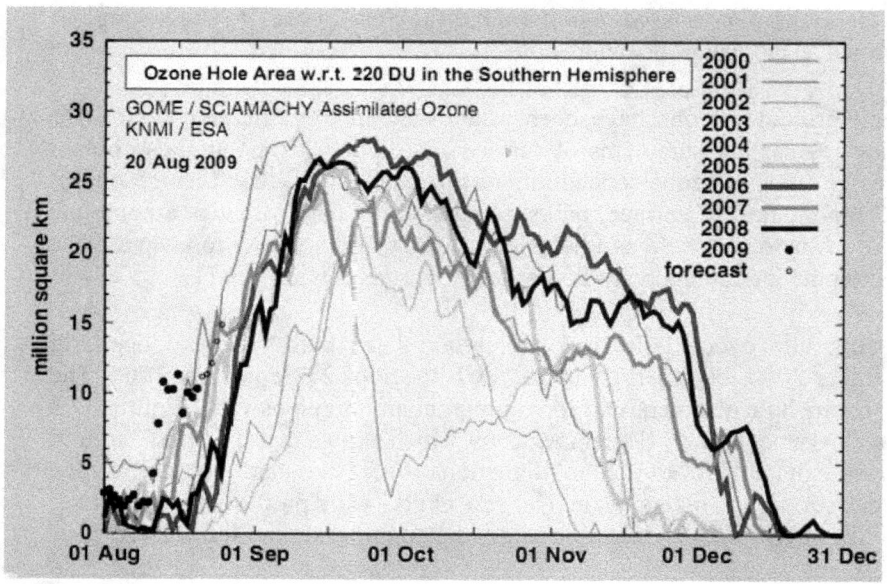

Figure 1.7.1 Interannual variation of the area of the Antarctic ozone hole between August and December during 2000-2008 based upon data from GOME and SCIAMACHY satellite instruments (Source: WMO 2009b)

Since chlorine and bromine gases play a crucial role in the formation of the ozone hole, and these come from chemicals like chloro-fluoro-carbons (CFCs), the production of such chemicals have been banned under the Montreal Protocol of 1987 and other international agreements. Recent observations and several studies have shown that the size of the annual ozone hole has stabilized and the level of ozone-depleting substances has decreased by 4 % since 2001. However, since chlorine and bromine compounds have long lifetimes in the atmosphere, a recovery of atmospheric ozone is not likely to be noticeable until 2020 or later.

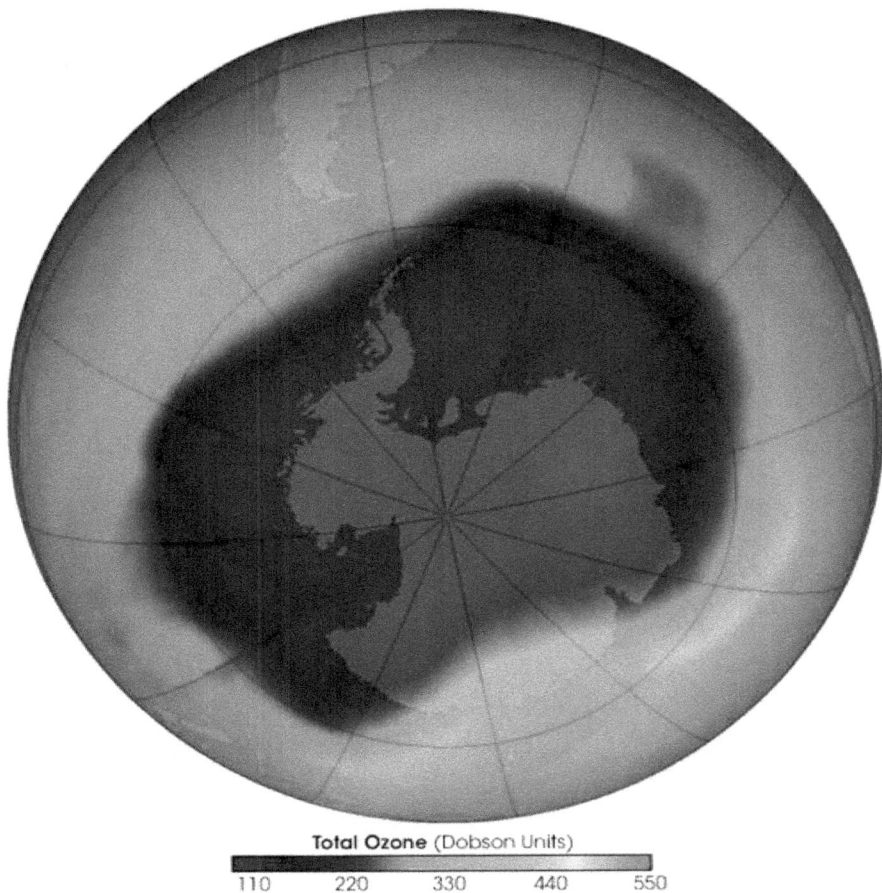

Figure 1.7.2 Antarctic ozone hole as on 24 September 2006
(Source: NASA web site earthobservatory.nasa.gov/IOTD)

1.7.1 Ozone over the Indian Region

Decades before the ozone hole had been discovered, there had already been a considerable scientific interest in the study and measurement of atmospheric ozone. The first ozone measurements were made in India in 1928 at Kodaikanal, and in the 1840s, extensive work was done by Professor K. R. Ramanathan and his colleagues which gained international recognition (Ramanathan et al 1957, Kulkarni et al 1959, Karandikar 1952). India is fortunate to have continuous measurements of surface ozone amounts, total column ozone and upper atmospheric ozone profiles at several locations in the country since then. Under the auspices of WMO, an extensive global network of spectrophotometer stations has been established where ground-based measurements of total column ozone and surface ozone levels are made. At some locations, vertical ozone profiles are being obtained with ozonesondes, but the availability of data is confined to land regions.

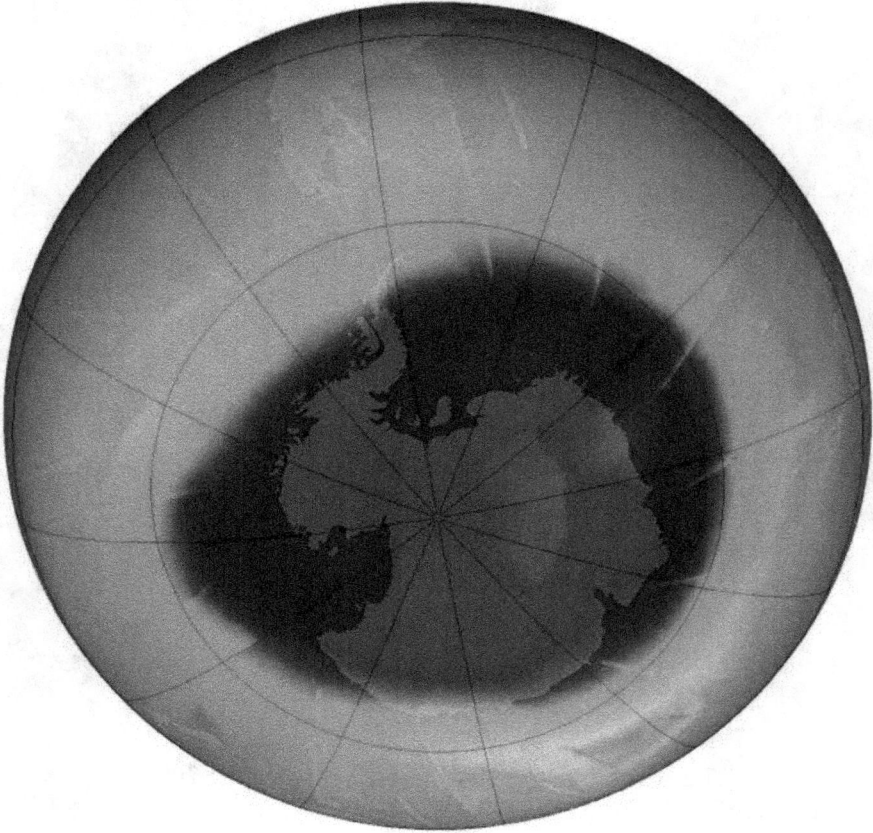

Figure 1.7.3 Antarctic ozone hole as of 10 September 2009
(Source: NASA web site earthobservatory.nasa.gov/IOTD)

After the discovery of the Antarctic ozone hole, there has been a revival of interest in ozone concentrations over the Indian region. Patil et al (2009) have used daily total ozone data from TOMS on Nimbus-7 (1979–1993) and Earth Probe (1997–2005) satellites to derive the characteristic features of extremes in the total ozone content and the frequency of the low/high ozone days have been carried out over the northern parts of India in the winter season. They observed that there was an increasing trend in the frequency of low ozone days and a decreasing trend in the frequency of high ozone days during the 1979-1993 period. Maximum as well as mean total ozone during January and February for this period shows a decreasing trend. However, the trends were not statistically significant over the more recent period 1997-2005. Revadekar et al (2009) in a parallel study using similar data have addressed the question whether or how far the atmosphere is on the path to ozone recovery due to the implementation of the Montreal Protocol and its amendments. Trends in extremes in the total ozone content over northern India during the recent period 1997-2005 compared to the earlier period 1979-1988 in general indicate some recovery but most of the trends are statistically insignificant.

Debaje et al (2009) have made a study of the variability of surface ozone over five different sites in western Maharashtra in India during the years 2001-2005. They found a pronounced maximum concentration of about 40-50 ppbv in the summer and winter seasons in both urban and rural sites despite lesser emission of precursor gases at the rural sites. An increase in ozone concentration was observed in the premises of a sugar factory during the sugarcane crushing period.

Fadnavis et al (2009) have investigated the spatiotemporal variations of the quasi-biennial oscillation (QBO) in temperature and ozone over the tropical and subtropical belts using Microwave Limb Sounder data for the period 1992-1999. Wavelet analysis was performed to study inter-annual variations in amplitude and phases of the QBO. Latitude-height cross sections of the amplitudes of temperature and ozone QBO exhibited a double peak structure near the equator.

Ali et al (2009) have reported that surface ozone over the Arabian Sea during the southwest monsoon season of 2002 showed an unusually low level of surface ozone with an overall average of 9 nmol/mol. They have given a novel explanation of this observation in terms of the destruction of ozone by reactive halides released from sea salt aerosols.

1.8 References

Ali K. and coauthors, 2009, "Sink mechanism for significantly low level of ozone over the Arabian Sea during monsoon", *J. Geophysical Research*, 114, D17306, doi:10.1029/2008JD011256.

Arrhenius S., 1896, "On the influence of carbonic acid in the air upon the temperature of the ground", *Philosophical Magazine*, 41, 237-276.

Callendar G. S., 1938, "The artificial production of carbon dioxide and its influence on temperature", *Quarterly Journal Royal Meteorological Society*, 64, 223-237.

Callendar G. S., 1949, "Can carbon dioxide influence climate?", *Weather*, 4, 310-314.

Carslaw K. S., Harrison R. G. and Kirkby J., 2002, "Cosmic rays, clouds, and climate", *Science*, 298, 1372-1377.

Chamberlain T. C., 1899, "A group of hypotheses bearing on climatic changes", *J. Geology*, 5, 653-663.

Debaje S. B. and Kakade A. D., 2009, "Surface ozone variability over western Maharashtra, India", *Journal of Hazardous Materials*, 161, 686-700.

Fadnavis S. and Beig G., 2009, "Quasi-biennial oscillation in ozone and temperature over tropics", *J. Atmospheric and Solar-Terrestrial Physics*. 71, 257-263.

Hansen J. and coauthors, 1981, "Climate impact of increasing atmospheric carbon dioxide." *Science*, 213, 957-966.

Karandikar R. V., 1952, "Measurements of atmospheric ozone and its vertical distribution at Kodaikanal (Lat. 10N)", *Proc. Indian Academy Sciences*, 35A, 290-302.

Kelkar R. R., 1970, "Role of Radiation in Atmospheric Circulation", *Ph. D. Thesis*, University of Poona, Pune, India, 241 pp.

Kiehl J. and Trenberth K., 1997, "Earth's annual global mean energy budget", *Bulletin American Meteorological Society.*, 78, 197-206.

Koppen W., 1900, "Versuch einer klassifikation der klimate, vorzugsweise nach ihren beziehungen zur pflanzenwelt", *Geogr. Zeitschr.* 6, 593-611, 657-679.

Kulkarni R. N., Angreji P. D. and Ramanathan K. R., 1959, "Comparison of ozone amounts at Delhi, Srinagar and Tateno in 1957-58", *Papers Meteorology Geophysics*, Meteorological Research Institute, Tokyo, 85-92.

Le Treut H. and coauthors, 2007, "Historical Overview of Climate Change", *Climate Change 2007: The Physical Science Basis. Contribution of Working Group I to the Fourth Assessment Report of the Intergovernmental Panel on Climate Change* [Ed: Solomon S. et al], Cambridge University Press, 94-127.

Manabe S. and Wetherald R. T., 1967, "Thermal equilibrium of the atmosphere with a given distribution of relative humidity", *J. Atmospheric Science*, 24, 241-249.

Patil S. D. and Revadekar J. V., 2009, "Extremes in total ozone content over northern India", *International Journal of Remote Sensing*, 30, 2389-2397.

Pierce J. R. and Adams P. J., 2009, "Can cosmic rays affect cloud condensation nuclei by altering new particle formation rates?, *Geophysical Research Letters*, 36, L09820, doi:10.1029/2009GL037946.

Plass G. N., 1956, "The carbon dioxide theory of climatic change", *Tellus*, 8, 140-154.

Ramanathan K. R. and Dave J. V., 1957, "The calculation of the vertical distribution of ozone by Gotz-Umkehr effect, method B", *Annals International Geophysical Year*, 5, 23-45.

Revadekar J. V. and Patil S. D., 2009, "Recent extremes in total ozone content over the northern parts of India in view of the Montreal Protocol", *International Journal of Remote Sensing*, 30, 3967-3974.

Solomon S. and coauthors (Ed), 2007, *Technical Summary, Climate Change 2007: The Physical Science Basis. Contribution of Working Group I to the Fourth Assessment Report of the Intergovernmental Panel on Climate Change*, Cambridge University Press, 91 pp.

Thornthwaite C. W., 1948, "An approach toward a rational classification of climate", *Geographical Review*, 38, 55-94.

Trenberth K. E., Fasullo J. T. and Kiehl J., 2009, "Earth's global energy budget", *Bulletin American Meteorological Society*, 80, 311-324.

Trewartha G. T., 1943, *An Introduction to Weather and Climate*, McGraw Hill, New York, 545 pp.

Trewartha G. T., 1961, *The Earth's Problem Climates*, University of Wisconsin Press, Madison,

WMO, 2009a, *WMO Greenhouse Gas Bulletin No.5*, World Meteorological Organization, Geneva, 4 pp.

WMO, 2009b, *WMO Antarctic Ozone Bulletin 1/2009*, World Meteorological Organization, Geneva, 17 pp.

Zillman J. W., 2009, "A history of climate activities", *WMO Bulletin*, 58, 141-150.

Chapter 2

Climate Monitoring

The Intergovernmental Panel on Climate Change (IPCC) which periodically reviews the state of the world's climate, does not itself monitor global climate or climate change and all its assessments are based on data collected from other sources and scientific studies made by others. In 1992, four years after the IPCC came into being, the Global Climate Observing System (GCOS) was established under the co-sponsorship of the World Meteorological Organization (WMO), the Intergovernmental Oceanographic Commission (IOC), the United Nations Environment Programme (UNEP), and the International Council for Science (ICSU). It is the objective of GCOS to ensure that the observations and information needed to address climate-related issues are obtained and made available to all potential users.

GCOS is a long-term, user-driven operational system capable of providing comprehensive observations required for monitoring the climate system, for detecting and attributing climate change, for assessing the impacts of climate variability and change, and for supporting research towards improved understanding, modelling and prediction of the climate system. GCOS addresses the total climate system including physical, chemical and biological properties, and atmospheric, oceanic, terrestrial, hydrological and cryospheric components.

Again, the GCOS programme does not directly make observations nor generate data products on its own. It only provides an operational framework for integrating observations made by participating countries and organizations. It builds upon, and works in partnership with, other existing and developing observing systems such as the WMO Global Observing System (GOS) the Global Atmosphere Watch (GAW), the Global Ocean Observing System (GOOS), and the Global Terrestrial Observing System (GTOS) and other in situ, airborne and space-based observational systems.

In this chapter, we will discuss how climate and climate change are monitored through various international cooperative efforts, and why it is not easy to obtain a reliable and error-free data series of even the basic climate-related parameters in spite of the satellites and other sophisticated technology tools that are available.

2.1 Surface Air Temperature

The temperature of the earth is clearly the most fundamental of all climate parameters and it needs to be monitored with great precision in order to get an accurate estimate of global warming.

There are currently 11,000 stations on land that make regular 3-hourly or hourly observations near the earth's surface of meteorological parameters such as atmospheric pressure, precipitation, wind speed and direction, air temperature and relative humidity. The stations are maintained by the respective national meteorological services, but they form a part of the WMO Global Observing System (GOS). A glance at the surface observatory network (Figure 2.1.1) is enough to bring to one's notice the uneven distribution of the stations over various parts the globe. The density of stations over Europe stands in vivid contrast with the sparseness of the network over many parts of Africa, Asia and south America. The network is very thinly spread over mountains, deserts and other uninhabited regions of the world, and absent over the oceans.

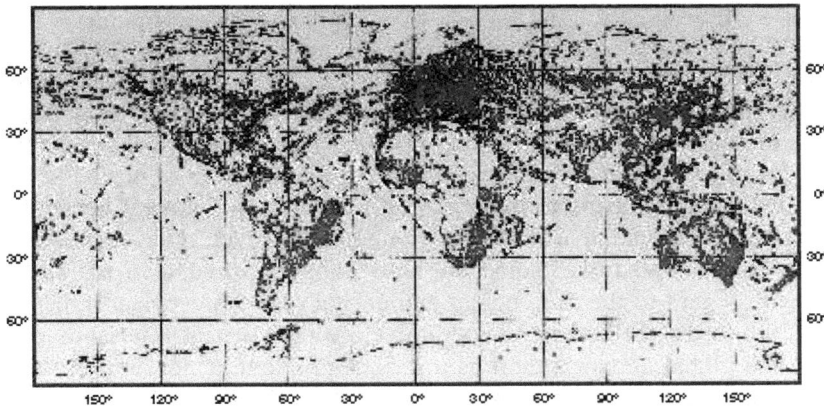

Figure 2.1.1 Network of 11,000 surface stations under the WMO Global Observing System (Source: WMO web site www.wmo.int)

Data from many of these 11,000 surface stations do not go into climatological analysis for various reasons like the short length of their data records or close proximity between stations. Only about 4,000 stations comprise the Regional Basic Synoptic Network (RBSN) and about 3,000 stations comprise the Regional Basic Climatological Network (RBCN) of the WMO. Data from these stations are exchanged globally in real time. For climate monitoring purposes, a yet smaller subset of about 1,000 carefully chosen surface stations is used in the Global Climate Observing System (GCOS) Surface Network (Figure 2.1.2).

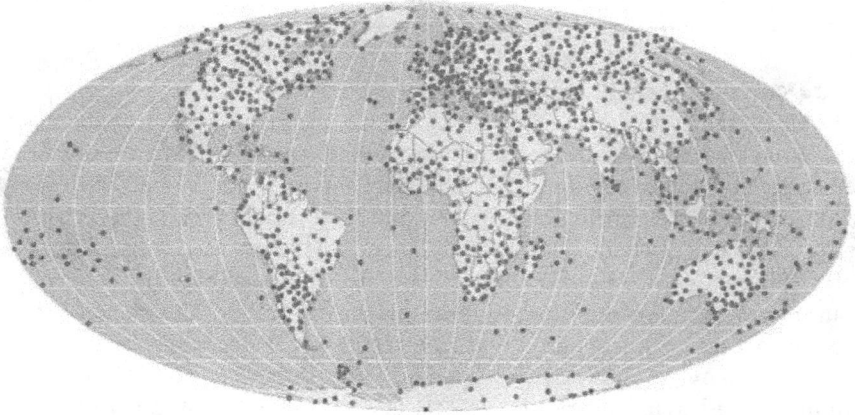

Figure 2.1.2 Network of 1,000 surface stations that comprise the Global Climate Observing System (Source: WMO web site www.wmo.int)

As an observational practice, it is recommended that air temperature at the surface be measured with a thermometer installed in a Stevenson screen at a height of 1.4 m (4 feet) above the ground. If the thermometer is placed at a different height, the temperature recorded will be slightly different because of the convective and radiative processes around it. However, many countries have in the past adopted, and some are still following, their own national practices with regard to the size, height and design of the Stevenson screen. Therefore, while the surface air temperatures are recorded at thousands of stations in the global network, there is a serious problem of non-uniformity of observational practice and incompatibility of data. As one goes backwards in time in order to construct a continuous time series and work out climate statistics, the situation gets worse. For computing both spatial and temporal averages, temperatures recorded at different places and in different years have to be normalized, but this may not always be possible or done.

Several issues have to be taken into consideration while dealing with historical surface air temperature data. Perhaps, the thermometers may not have been properly calibrated, uniform observational practices may not have been followed, proper exposure conditions may not have been ensured, stations might have been moved from one location to another without a record being kept, and so on. Historical temperature records have therefore to be very strictly quality controlled and compatibility of different data sources ensured while constructing long series of globally averaged temperatures required for climate change investigations.

2.2 Sea Surface Temperature

The oceans together occupy about two-thirds of the earth's surface area and they play a major role in determining the earth's weather and climate, but there are no observatories on the oceans like on land, although there are stations on some islands, Temperature, salinity and density are the three basic characteristics of ocean water. Moreover, unlike the atmosphere, the ocean is horizontally stratified, and these basic ocean characteristics vary much more vertically within the ocean than across it. The temperature could possibly drop by 5 °C from the sea surface to a depth of 1 km, but the horizontal temperature gradient may be of the order of only 5 °C in 5000 km. In situ measurements of ocean parameters are very important but it is extremely difficult to cover the vast expanse of the oceans, that too in three dimensions. The cost of operating and maintaining meteorological and oceanographic research vessels is high, and only a few countries can afford to have them.

Much of the Sea Surface Temperature (SST) climatology that we have today is based upon observations by the Voluntary Observing Fleet (VOF) of merchant navy ships. These ships undertake to carry out the additional responsibility of reporting weather observations and keeping weather logs, but a major constraint in this effort is that such observations are confined to narrow designated shipping lanes in which the merchant vessels sail. As a result, vast areas of oceans still continue to remain uncharted. SST has traditionally been measured by ships by the bucket method, in which a bucket is lowered down from the deck, allowed to get filled with sea water and pulled up. A thermometer is placed in the water sample collected and the measured temperature is noted. The SST so recorded represents the temperature of the sea water at a depth of approximately 1 m below the surface.

Observations made by ships recruited under the WMO VOF Programme, comprise much the same variables as at surface land stations with the important additions of sea surface temperature, wave height and period. The number of voluntary observing ships is currently around 4,000 and about 1,000 of them report observations every day. However, ship observations are confined to merchant shipping lanes and coverage over the southern hemisphere is relatively quite poor.

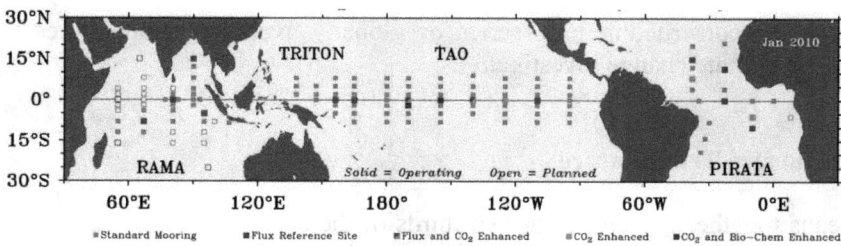

Figure 2.2.1 Array of 200 moored buoys in the tropical oceans
(Source: NOAA web site www.aoml.noaa.gov)

Under several international programmes, moored ocean buoys have been
installed in various tropical ocean regions and they presently number about
200 (Figure 2.2.1). They record SST and several surface and subsurface
ocean parameters and transmit them via satellite. The moored buoys are
complemented by drifter buoys in the global oceans which are not static but
keep moving along with the ocean currents. As of January 2010 there were
1,200 drifter buoys in the global oceans which record and transmit SST data
via satellite to a receiving centre (Figure 2.2.2). These buoys provide over
27,000 SST measurements per day and a half of the drifters also report sea
level pressure providing about 14,000 reports per day.

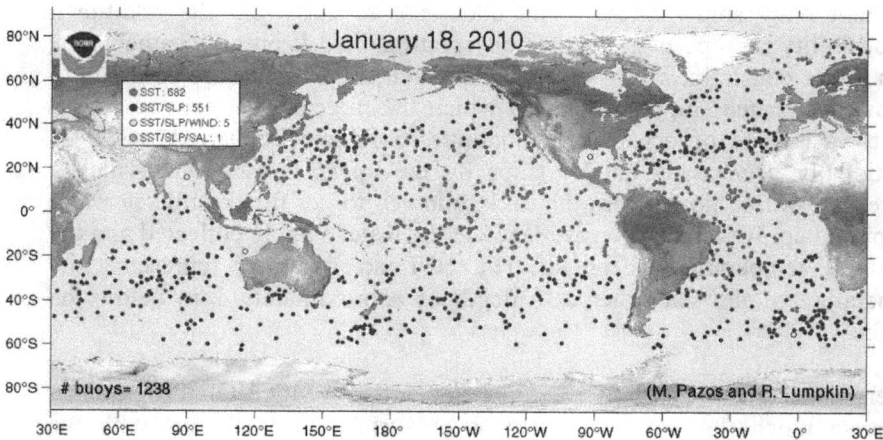

Figure 2.2.2 Array of 1,200 drifter buoys in the global oceans
as of January 2010 (Source: NOAA web site www.pmel.noaa.gov)

Argo is an international programme launched in 2002 and is aimed at
obtaining high quality temperature and salinity profiles within the upper
2,000 m of the oceans. It is supported by as many as 23 different countries.

As of December 2009, the Argo array consisted of more than 3,100 Argo floats (Figure 2.2.3), about 160 of which have been contributed by India and deployed over the Indian Ocean. About a half of the total number of Argo floats have been provided by the U. S. alone.

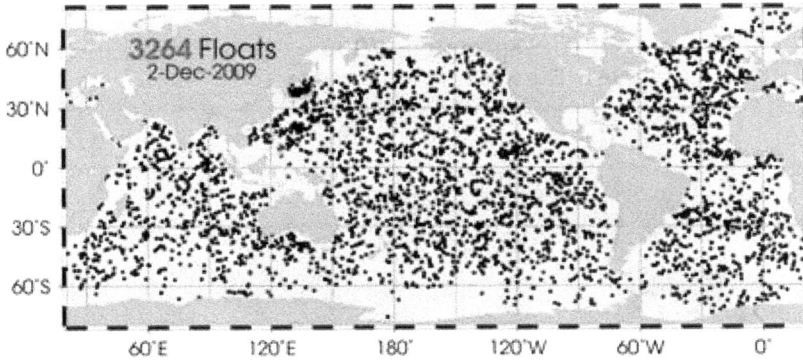

Figure 2.2.3 Network of 3,200 Argo floats as of December 2009
(Source: Argo web site www.argo.net)

The Argo float records temperature and salinity profiles of the ocean up to a depth of 2 km. The float is battery-powered and remains mostly at this depth, but once every 10 days, it is made to rise to the ocean surface. During its ascent it makes about 200 measurements of pressure, temperature and salinity which are stored on board. When the float reaches the ocean surface, it remains there for 6 to 12 hours during which it transmits the stored data to a ground station via satellite. It then returns to its parking depth. A float is designed to make about 150 such ascent-descent cycles during its lifetime.

The Argo data are going to provide a quantitative description of the changing state of the upper ocean and the patterns of ocean climate variability from months to decades, including heat and fresh water storage and transport. They complement satellite altimeter data on sea surface height. Assimilation of Argo data is valuable for initializing ocean and coupled ocean-atmosphere models.

It was only with the advent of satellite remote sensing in the 1970s that it came possible to monitor the global oceans extensively and continuously. A satellite radiometer measures radiation over finite wavelength bands. The radiance measured has an associated brightness temperature, or the temperature at which a black body would emit the same radiation. From this the body's true temperature, that is the earth's surface temperature, can be estimated if its emissivity is known. As the emissivity of the sea surface can be safely assumed as a very close approximation to have a value of 1, the satellite-measured thermal infrared window radiance can be inverted to get the temperature of the sea surface. This is the basic principle of satellite-based SST retrievals.

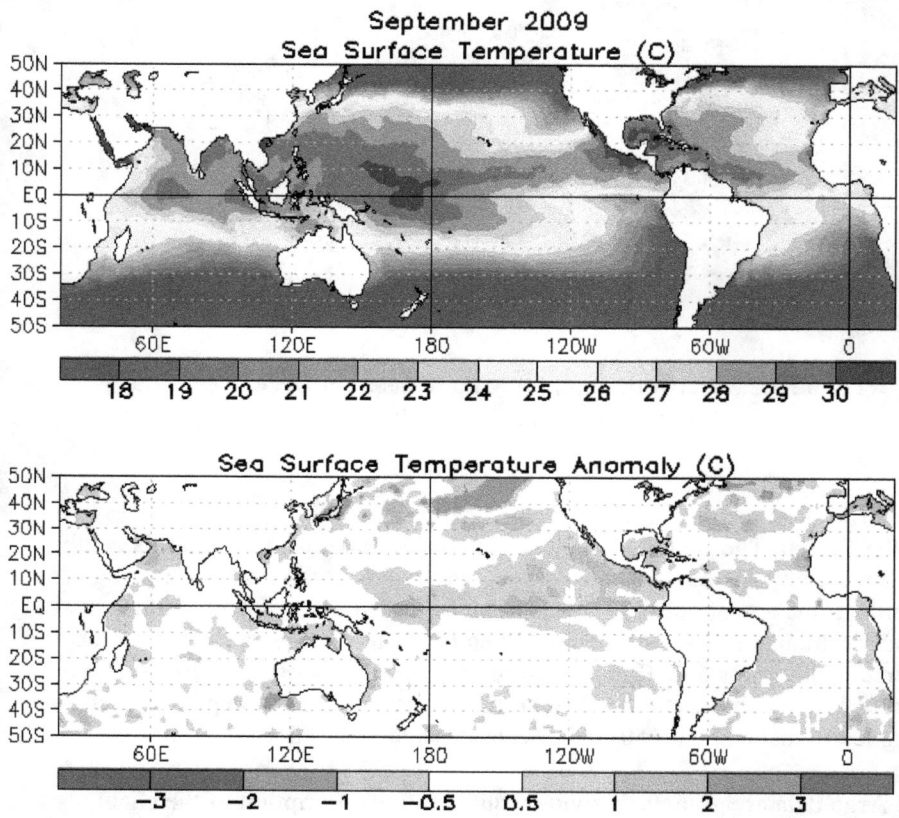

Figure 2.2.4 Mean (top) and anomalous (bottom) SST for September 2009.
Anomalies are departures from the 1971-2000 mean.
(Source: NOAA web site cpc.noaa.gov)

Satellite-derived SSTs represent the skin temperature of the ocean as they are retrieved by an inversion of the radiative transfer equation, assuming the sea surface to be a black body. In practice, satellite-derived SSTs are vitiated by the moisture that resides in the atmosphere and attenuates the emission from the sea along its path before reaching the satellite. This has to be corrected for by using multiple infrared window radiances and applying suitable algorithms. In the presence of clouds, SST retrievals are very difficult or even impossible to make as the brightness temperature would represent the cloud top temperature. SST retrievals from passive microwave measurements overcome the clouding problem but suffer from degraded spatial resolution. This is due to the inherent problem of the low energy associated with radiation in the microwave region of the spectrum compared with that in the thermal infrared window region.

Global SST charts and SST anomaly charts are being prepared and disseminated by agencies like the U. S. National Oceanic and Atmospheric Administration (NOAA). Figure 2.2.4 is an example of a monthly SST charts. However, different kinds of users have different requirements of resolution and accuracy. For operational numerical weather prediction, an accuracy of 0.2-0.5 °C at 100 sq km and 3-day resolution is required. For ENSO prediction, 0.2-0.3 °C accuracy at 30-100 sq km and 5-day resolution are adequate. For mesoscale and coastal oceanography purposes, it has to be 0.2 °C on a 10 sq km scale. However, for climate change investigations, a SST accuracy of 0.1 °C is expected although a 500 sq km resolution on a monthly scale is sufficient.

To summarize, SST measurements have traditionally been made by ships using the bucket method and these date back to 1850 or so. The long period of ship data is an asset for climate change studies, but the data is confined to merchant shipping lanes. In situ SST measurements made by moored and drifting ocean data buoys are available for the last 20 years or so. The Argo float programme is the most recent one and the data generated is not yet adequate for climate scale processing. The extensive and continuous SST data retrieved from satellites goes back to the 1970s but these are ocean skin temperatures. Thus the main problem in using SST data is its heterogeneity with respect to time, space, accuracy and depth of measurement. A lot of effort is therefore required to be put in for blending them together and normalizing them in order to build a unified data series accurate enough for detecting climate change signals.

2.3 Computing the Global Average Surface Temperature

The first problem in uniquely computing the earth's average surface temperature is that we are dealing with data from thousands of stations over the last 150 years, over which there have been changes in their total number, location, exposure conditions, calibration, observation techniques and so on. Then, measurements made over land and sea are intrinsically different, and there are no observatories over the oceans, which occupy two-thirds of the earth's surface area, and ship observations are confined to commercial shipping lanes only. There are data gaps over large uninhabited land regions in conventional measurements which can be filled in by satellite-derived retrievals, but the satellite data series are of a much shorter length and have their own errors, uncertainties and incompatibilities. Like in Figure 2.2.4, the SST anomalies are computed with respect to the average over only a 30-year period 1971-2000 for which retrievals are available.

In essence, we are dealing basically with an assortment of surface temperature data, and depending upon what techniques we adopt to bring in compatibility within it, we can derive different values of the global average temperature of the earth. Further, we can derive different values of the year-to-year temperature anomalies depending upon what average value we choose. This is not simply a hypothetical argument, but it is the actual situation.

In fact, the extent of global warming is being monitored in parallel by three independent agencies. One is the U. K. Met Office Hadley Centre at Exeter working together with the Climatic Research Unit of the University of East Anglia (HadCRUT). Another is the National Climatic Data Center (NCDC) of the U. S. National Oceanic and Atmospheric Administration (NOAA) in Washington D. C. The third is the Goddard Institute for Space Studies (GISS) of the U. S. National Aeronautics and Space Administration (NASA), in New York. These three agencies maintain three different data bases on the earth's surface temperature and come up every year with their own individual results and statistics on global warming.

The HadCRUT data series uses 1961-1990 as the 30-year period of averaging that covers more recent warm years. GISS uses an earlier 30-year averaging period of 1951-1980. NCDC has a much longer averaging period 1901-2000 that covers the entire 20th century. The earth's average surface temperature therefore has slightly different values in the three different data sets and the annual temperature anomalies derived with respect to this average are also different (Figure 2.3.1 to 2.3.3). There are different rankings of the warmest years as well.

In order to overcome the data problems that have been mentioned in the preceding sections, it has become a common practice to describe global warming not in terms of the absolute values of global average surface temperature but in terms of their anomalies or departures from a long-term mean value. HadCRUT, NCDC and GISS, have all adopted this practice for their global temperature analyses.

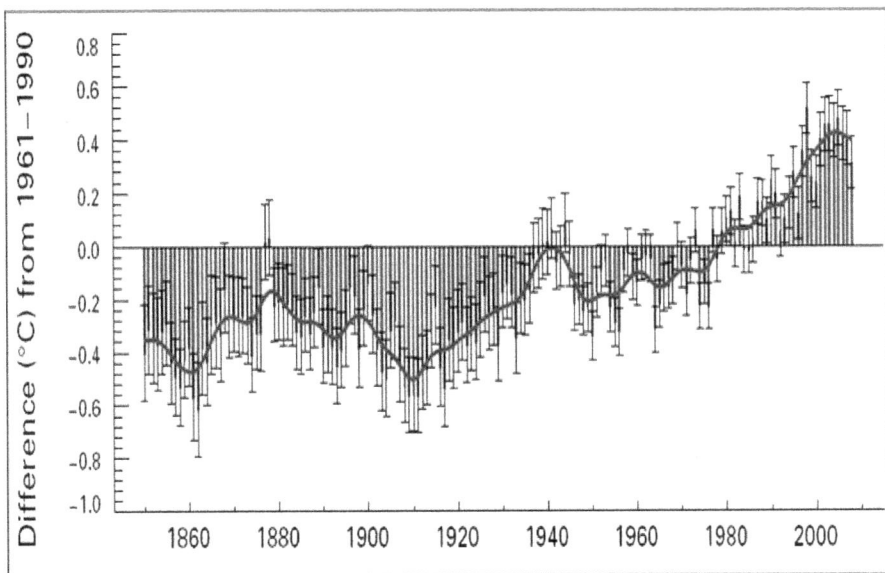

Figure 2.3.1 HadCRUT annual average global surface temperature anomalies
(°C) for the period 1850-2008, relative to the average
for 1961-1990 (Source: WMO web site www.wmo.int)

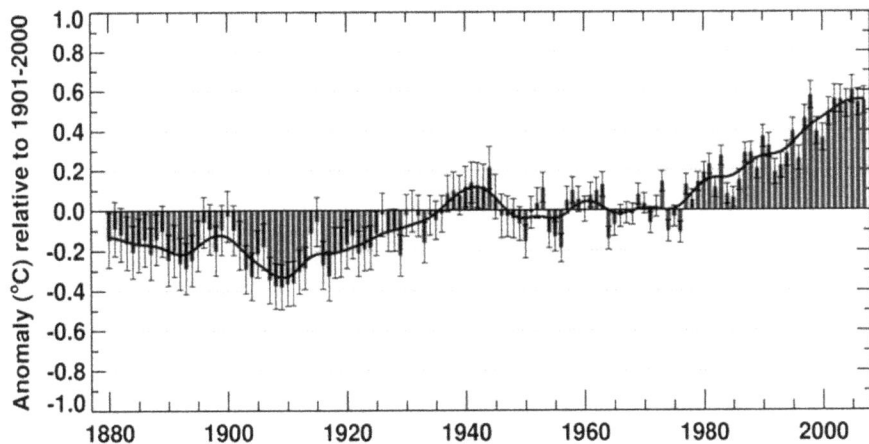

Figure 2.3.2 NCDC annual global surface temperature anomalies (°C) for the
period 1880-2009, relative to the average for 1901-2000
(Source: NCDC web site www.ncdc.noaa.gov.in)

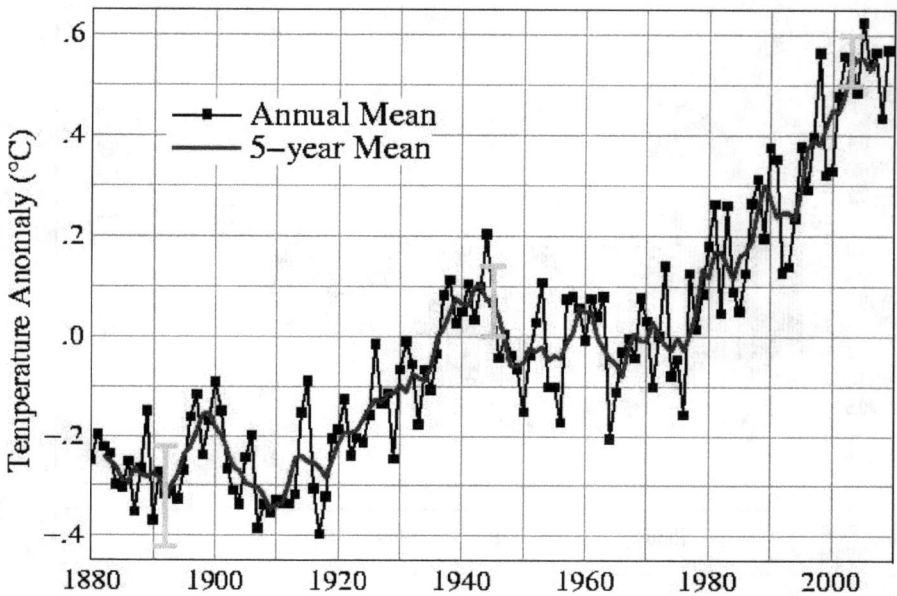

Figure 2.3.3 GISS annual global surface temperature anomalies (°C) for the period 1880-2009, relative to the average for 1951-1980. The dotted black line is the annual mean and the solid red line is the 5-year mean. The green bars show uncertainty estimates.
(Source: GISS web site http://giss.nasa.gov.in)

Anomalies have been found to describe climate variability over larger areas more accurately and realistically than absolute temperatures do, and they give a frame of reference that allows more meaningful comparisons between locations and more accurate calculations of temperature trends. Absolute surface temperature can change markedly even over a short distance, particularly because of topographic variations, while temperature anomalies can be similar over much larger regions. Temperature anomalies are strongly correlated among themselves even over distances of the order of 1000 km.

The rationale for using global temperature anomalies instead of absolutes can be justified on several counts. Some regions have few temperature measurement stations necessitating interpolation over the intervening data gap regions, which is difficult when they are not homogeneous. In mountainous areas, most observations come from the inhabited valleys, so the effect of elevation on a region's average temperature must be considered as well. For example, a summer month over an area may be cooler than average, both at a mountain top and in a nearby valley, but the absolute temperatures will be quite different at the two locations. The use of anomalies in this case will show that temperatures for both locations were below average.

The land component of the Hadley Centre global temperature data set (HadCRUT) is derived from a collection of homogenized, quality-controlled, monthly average temperatures for over 4,000 stations. Brohan et al (2006) have discussed in detail the practical problems that are encountered with the compilation of historical climate data sets. The uncertainties in the land data can be divided into three groups: (1) station errors arising from problems associated with individual stations, (2) sampling errors arising from the averaging of temperatures over a grid box with insufficient number of stations within it, (3) bias errors resulting from systematic changes in measurement methods, exposure changes and urbanization of the area in which a station is located.

The ocean data set also has its own errors like measurement and sampling errors and bias errors. An additional source of uncertainty arises from the fact that unlike land stations which have permanent locations, point measurements over the sea are made on moving platforms like ships and buoys. Hence the configuration of point data sources within a grid box is not fixed but varies with time.

Finally to make a global data set, the land and marine data must be combined. For land-only grid boxes the land value is taken, and for sea-only grid boxes the marine value, but for coastal and island grid boxes the land and marine data is blended into a combined average.

The time series of global surface temperature is produced by NCDC from the blended land and ocean data set of Smith and Reynolds (2005). It consists of monthly average temperature anomalies on a 5° x 5° grid across land and ocean surfaces. These grid boxes are then averaged to provide an area-weighted average global temperature anomaly. Global average anomalies are calculated on a monthly and annual time scale. Average temperature anomalies are also available for land and ocean surfaces separately, and the northern and southern hemispheres separately. The global and hemispheric anomalies are provided with respect to the average for the period 1901-2000, which has the widest distribution for historical data. The absolute values of global monthly surface temperatures are given in Table 2.3.1 for different months and for land and sea separately. As per NCDC, the earth's average surface temperature for the period 1901-2000 is 13.9 °C.

Table 2.3.1 Global Monthly Surface Temperature averaged for the Period 1901-2000
(Source: NCDC web site www.ncdc.noaa.gov.in)

Month	Mean Temperature (°C)		
	Land	Sea	Global (Land-Sea Combined)
Jan	2.8	15.8	12.0
Feb	3.2	15.9	12.1
Mar	5.0	15.9	12.7
Apr	8.1	16.0	13.7
May	11.1	16.3	14.8
Jun	13.3	16.4	15.5
Jul	14.3	16.4	15.8
Aug	13.8	16.4	15.6
Sep	12.0	16.2	15.0
Oct	9.3	15.9	14.0
Nov	5.9	15.8	12.9
Dec	3.7	15.7	12.2
Annual	8.5	16.1	13.9

The GISS analysis also concerns only temperature anomalies and not absolute temperature. Temperature anomalies are computed relative to the base period 1951-1980. The GISS method of deriving the average global surface temperature (Hansen et al 1981) has undergone several modifications over time (Hansen et al 2001). Historical surface temperature data have not all been recorded at identical hours, so corrections have been introduced for removing the biases due to time of observation bias. Station history adjustments have also been made to station data, particularly for the U. S., Canada and Mexico. Rural, small-town, and urban stations have been reclassified based upon satellite measurements of night light intensity in order to remove the potential effects of urbanization in such places on surface air temperature records. Hansen et al (2001) estimate the inherent uncertainties in the long-term temperature change at least of the order of 0.1°C for both the U. S. mean and the global mean. The successive periods of global warming (1900-1940), cooling (1940-1965), and warming (1965-2000) in the 20th century show distinctive patterns of temperature change suggestive of roles for both climate forcings and dynamical variability.

The different analysis techniques followed by HadCRUT, NCDC and GISS result in the top ten warmest years being ranked somewhat differently. As per the HadCRUT analysis, 1998 was by far the warmest year on record

(Figure 1.6.2). The temperature in the year 2008 was 0.31 °C above the 1961-1990 annual average of 14.0 °C, ranking 2008 as the tenth warmest year on record. The year 2009 is likely to rank as the fifth warmest year with an anomaly currently estimated at 0.44°C but final figures are not yet available. However, the decade of the 2000s (2000-2009) was warmer than the decade spanning the 1990s (1990-1999), which in turn was warmer than the 1980s (1980-1989).

As per NCDC, global land and ocean annual surface temperatures tied with 2006 as the fifth warmest on record, at 0.56 °C above the 20th century average. The 2000-2009 decade is the warmest on record, with an average global surface temperature of 0.54 °C above the 20th century average. The average for the 1990s was 0.36 °C. Table 2.3.2 shows the relative ranking of the top ten warmest years as per NCDC. According to this ranking 2005 is the warmest tear on record and 1998 the second warmest. This differs from the HadCRUT ranking which places 1998 at the top. However it must be seen that the ranking is made in terms of changes of the order of a hundredth of a degree in the annual temperature anomaly.

Table 2.3.2 Relative ranking of the top ten warmest years as per NCDC (Source: NCDC web site www.ncdc.noaa.gov.in)

Global Top Ten Warm Years	Global Temperature Anomaly °C
2005	0.62
1998	0.60
2003	0.58
2002	0.57
2009	0.56
2006	0.56
2007	0.55
2004	0.54
2001	0.52
2008	0.48
1997	0.48

According to GISS estimates, the year 2008 was the coolest year since 2000, and the ninth warmest year since 1880. The ten warmest years all occur within the 12-year period 1997-2008. The error bar in comparing recent years is estimated as 0.05 °C, so that 2008 could have been somewhere within the range from 7th to 10th warmest year in the record.

2.4 Greenhouse Gases

Measurements of climate-related parameters such as greenhouse gases, ozone, aerosols and reactive gases were being made at many sites around the world since long, some of them dating back to the 1920s. When the issue of climate change came to the forefront, the need was felt for coordinating such diverse efforts and bringing them under one umbrella, so that authentic data sets would become available for climate monitoring and research. This was achieved in 1989, when the World Meteorological Organization established the Global Atmosphere Watch (GAW) programme. GAW was built upon the Global Ozone Observing System (GO_3OS) and the Background Pollution Monitoring Network (BAPMON) system which were in existence at that time, and relevant components of GCOS, with an increased emphasis on quality assurance and global partnership. GAW measurements fall into six broad categories: greenhouse gases, ozone, UV radiation, aerosols, major reactive gases and precipitation chemistry.

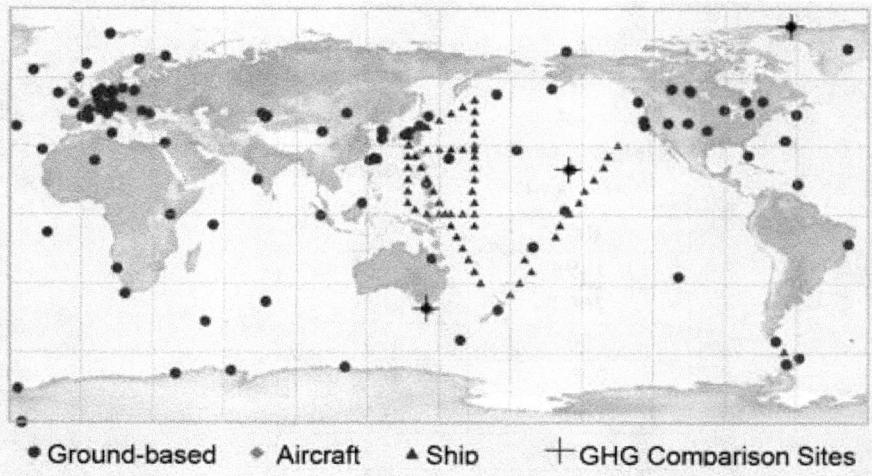

Figure 2.4.1 WMO/GAW global greenhouse gas network for measurement of carbon dioxide and methane concentrations. (Source: WMO 2009)

Sites where CO_2 and methane concentrations are monitored under the WMO/GAW programme are shown in Figure 2.4.1. Besides land and island stations, measurements are also made by ships and aircrafts. The measurement data are reported by participating countries and archived and distributed by the World Data Centre for Greenhouse Gases (WDCGG) at the

Japan Meteorological Agency (JMA). Statistics on the current global atmospheric abundances of the three major greenhouse gases as measured by the GAW stations and the increase in their concentrations since the industrial era have been given in Table 1.4.1. These values are slightly different from those in IPCC AR4, mainly due to the different selection of stations employed.

Accurate measurements of atmospheric CO_2 were first started in 1957 during the International Geophysical Year by David Keeling who chose the site at a height of 3,400 m on the slopes of the Mauna Loa volcano because he wanted to measure CO_2 in air masses that would be representative of the entire globe, at least of the northern hemisphere. The observatory at Mauna Loa, in Hawaii, continues to make these measurements uninterruptedly till now (Figure 2.4.2). For measurement of the CO_2 concentration in the atmosphere, air is slowly pumped through a small cylindrical cell with flat windows on both ends. Infrared radiation is transmitted through one window into the cell, and what emerges from the second window is measured by a detector that is sensitive to infrared radiation. The attenuation of radiation is a measure of the CO_2 amount in the atmospheric sample in the cell.

The quantity actually measured is the mole fraction, which is the number of carbon dioxide molecules in a given number of molecules of air, after removal of water vapour. It is expressed in parts per million (ppm) or the number of CO_2 molecules in a million molecules of dry air. The concentration of a gas in the atmosphere is defined as the number of molecules per cubic metre. However, it would depend upon the pressure and temperature, and the relative fraction of water vapour which is extremely variable. Hence the mole fraction of CO_2 in dry air is used to denote the increase or decrease of CO_2 in the atmosphere.

All of the measurements at Mauna Loa are rigorously and very frequently calibrated. Ongoing comparisons of independent measurements at the same site allow an estimate of the accuracy, which is generally better than 0.2 ppm. Before 1995 the calibration was being carried out by the Scripps Institution of Oceanography but since then the World Meteorological Organization (WMO) has taken over the responsibility of maintaining the CO_2 Central Calibration Laboratory and the WMO Mole Fraction Scale for CO_2-in-air. Details of the measurement and calibration procedures are available in the literature (Komhyr et al 1989, Zhao et al 2006).

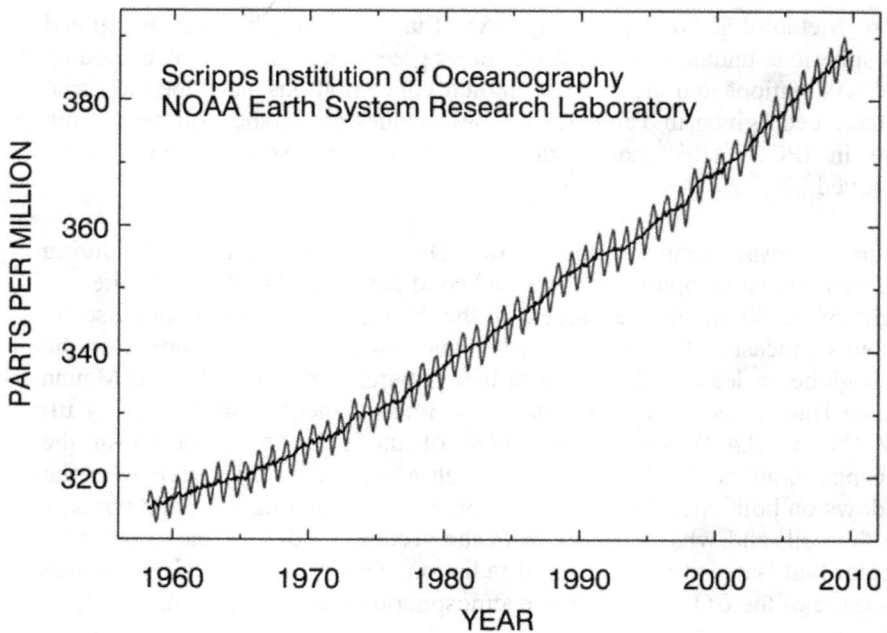

Figure 2.4.2 CO_2 concentration measured at Mauna Loa from 1958 to 2010
(Source: www.esrl.noaa.gov/gmd/ccgg/trends/co2_data_mlo)

2.5 Ozone

Measurements of total column ozone in the atmosphere have been regularly made at many stations around the world since the 1920s using Dobson spectrophotometers and since the 1980s with the new Brewer spectrophotometers. These instruments measure ultraviolet radiation from the sun at different wavelengths ranging between 305 to 345 nm. By measuring the ultraviolet radiation at two different wavelengths within this band, the amount of ozone can be estimated. At 305 nm wavelength, absorption by ozone is maximum, while the 325 nm wavelength is insensitive to the ozone in the atmospheric column. The ratio of the intensities in these two wavelengths provides a measure of the amount of ozone in the earth's atmosphere that the solar radiation has to traverse.

Ozonesondes are balloon-borne instruments which obtain the vertical profiles of ozone in the atmosphere by means of a chemical reaction. A pump is used to draw an air sample into a chamber that contains dilute potassium iodide which gets converted to iodine. There is another chamber which contains concentrated potassium iodide and the two chambers are connected by an ion bridge. The chemical reaction causes the two chambers of potassium iodide to fall out of equilibrium and electrons flow through a circuit creating an electrical current. The strength of the current is a measure of the ozone content in the air sample.

The Global Ozone Monitoring Network under the WMO/GAW programme consists of 132 stations that measure total column ozone with Dobson and Brewer spectrophotometers (Figure 2.5.1) and 63 stations where vertical profiles of ozone are obtained with balloon-borne ozonesondes (Figure 2.5.2). India has also been making regular measurements of total column ozone, surface ozone and ozone vertical profiles at several stations in India and also at its base stations in Antarctica.

The Network for the Detection of Atmospheric Composition Change (NDACC), set up in 1991, consists of more than 70 stations which make high-quality observations of the physical and chemical state of the stratosphere and the free troposphere for assessing the impact of atmospheric composition change on global climate. The Southern Hemisphere Additional Ozonesondes (SHADOZ) network was established in 1998 to make up for the scarcity of data over the southern hemispheric tropics and subtropics. It helps to correct the discrepancies in southern hemispheric ozone data, by coordinating and augmenting ozonesonde flights and centrally archiving the data. Currently, there are 14 active sites in the SHADOZ network.

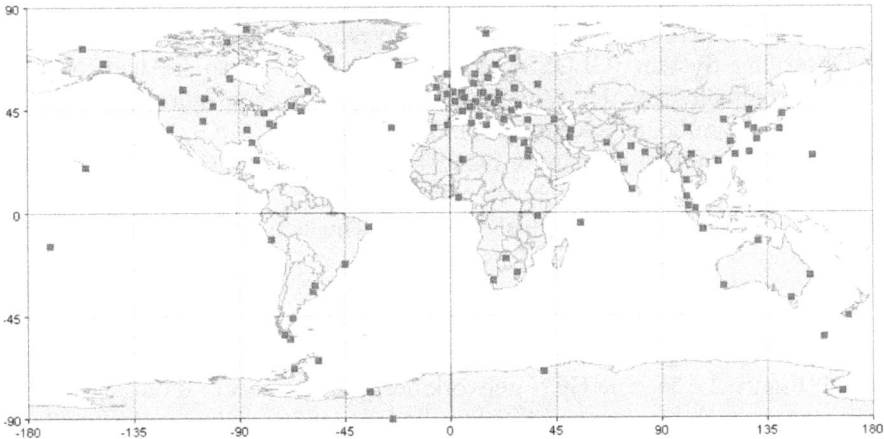

Figure 2.5.1 The WMO-GAW global network of Dobson/Brewer
spectrophotometers for measurement of total column ozone
(Source: WMO web site www.wmo.int)

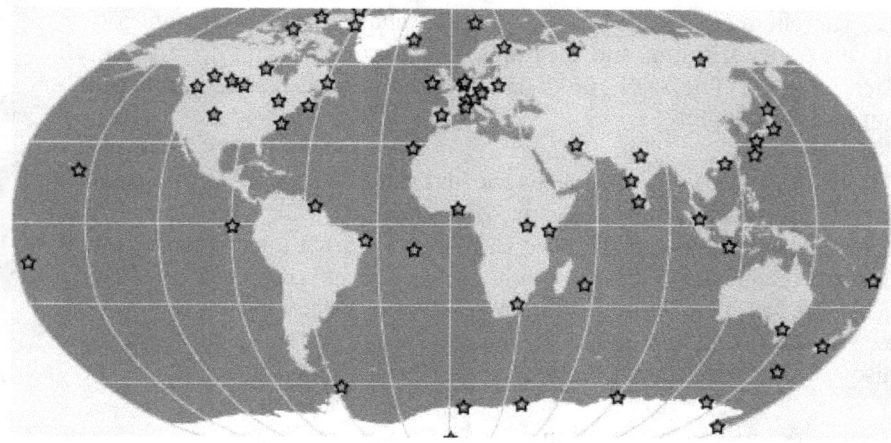

Figure 2.5.2 The combined network of ozonesonde stations for measurement of vertical profiles of ozone under WMO-GAW, SHADOZ and NDACC (Source: WMO web site www.wmo.int)

2.6 Sea Level

One of the repercussions of global warming that is likely to have disastrous consequence is a rise in the global sea level. It is therefore of vital necessity to monitor the sea level globally with an accuracy that is high enough to match the requirements of global warming investigations. The Global Sea Level Observing System (GLOSS), an international programme launched and jointly conducted by the World Meteorological Organization (WMO) and the Intergovernmental Oceanographic Commission (IOC), is of great importance and help in this regard. GLOSS aims at the establishment of high quality global and regional sea level networks for application to climate, oceanographic and coastal sea level research. The acronym GLOSS also stands for 'Global Level of the Sea Surface'.

The main component of GLOSS is what is called as the Global Core Network (GCN) which comprises 290 sea level monitoring stations around the world (Figure 2.6.1). The GCN network design is subject to modification every few years and the current GCN is called GLOSS02. GCN is designed to provide a sampling of coastal sea level variations that is distributed as evenly as possible around the globe. Another component of the GLOSS system is the Long Term Trends (LTT) set of gauge sites which are meant for monitoring long term trends and accelerations in global sea level. Some of the LTT stations are common to the GCN but not all. The LTT sites will have GPS receivers installed for monitoring vertical land movements, which can be removed from the relative sea level rise to obtain the absolute sea level rise required for climate change purposes.

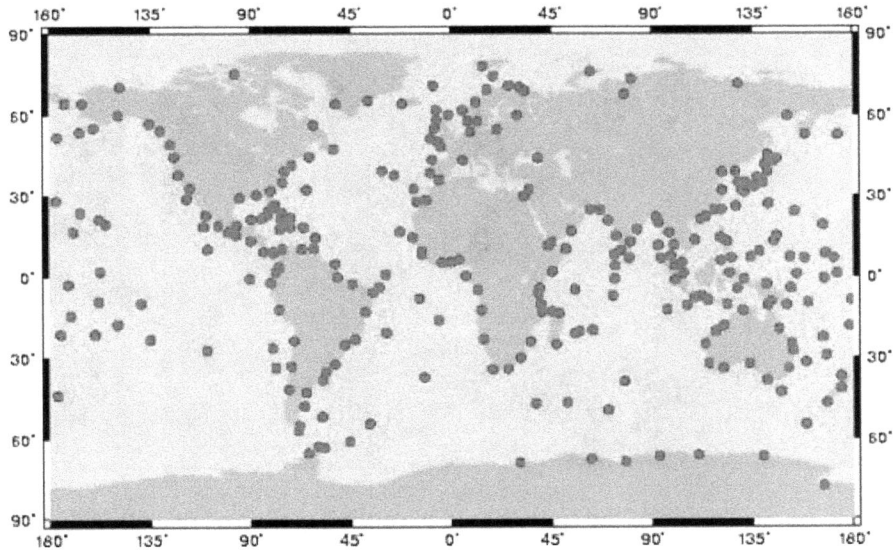

Figure 2.6.1 Global network of sea level observing stations
(Source: GLOSS web site gloss-sealevel.org)

2.7 Space-based Monitoring

Considering the problems associated with climate monitoring the global climate such as those discussed earlier (Section 2.3), and that it will take a long time to develop data series from the new in situ systems like Argo, it is clear that to get a complete global picture of the earth's climate, we need to build observing systems based on satellite remote sensing.

In the last fifty years after the launch of the world's first weather satellite TIROS-1 by the U. S, the global space observing system has evolved into a huge constellation of geostationary and orbiting satellites of different countries including India (Figure 2.7.1) Besides the many operational meteorological, environmental and land resource satellites which are in orbit, the U. S, in particular has launched several specific mission for monitoring parameters representing global climate change.

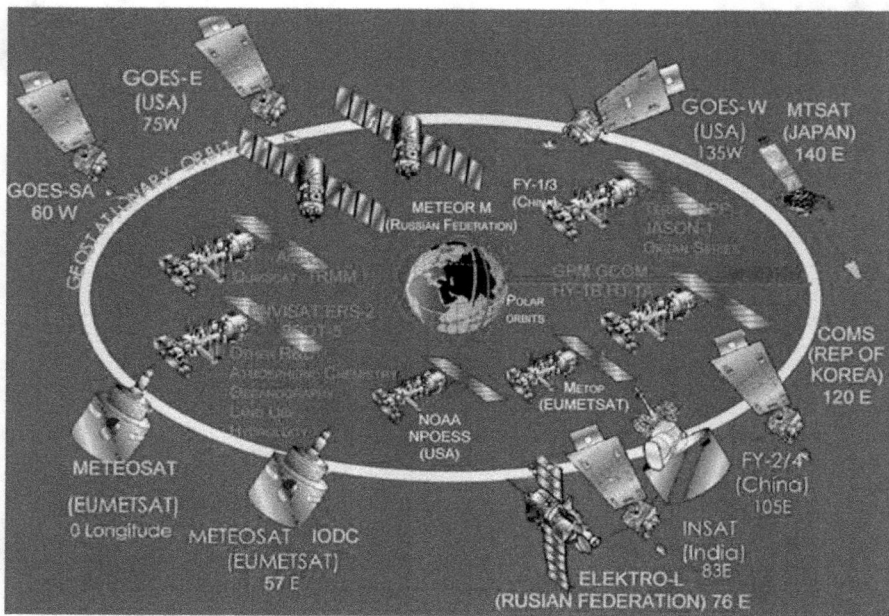

Figure 2.7.1 Schematic of the space-based observing system of operational
meteorological satellites of different countries \
(Source: WMO web site www.wmo.int)

The AcrimSat satellite launched in 1999, and the SORCE (Solar Radiation
and Climate Experiment) launched in January 2003, measure the solar energy
output with uniform sensitivity in the ultraviolet to infrared wavelengths.
Since 2006, the Calipso and CloudSat satellites have been complementing
each other's capability for measuring various properties of clouds and
aerosols in the atmosphere. The Tropical Rainfall Measuring Mission
(TRMM) has been a very successful joint U. S. - Japan mission and it is
going to be followed in 2013 by a very ambitious Global Precipitation
Mission in which many countries including India are going to participate.

Under the Gravity Recovery and Climate Experiment (GRACE) two
spacecrafts observe and measure the gravitational field of the earth from
which it is possible to derive the shape of the geoid and the distributions of
water and ice. ICESAT (Ice Cloud and Land Elevation Satellite) monitors
whether the size and thickness of the ice sheets are reducing because of
global warming. The Jason-1 spacecraft and its successor Ocean Surface
Topography Mission (OSTM) monitor the height of the sea surface globally
by radar altimetry.

The Aquarius satellite is planned for launch in 2010, and is expected to provide the first ever global maps of ocean salinity. Glory is a low earth orbit satellite that is also going to be launched in 2010, and is designed to collect data on the properties of aerosols, including black carbon, and also to study measure solar irradiance. The Soil Moisture Active and Passive (SMAP) satellite will provide global measurements of soil moisture and the state of the soil which is linked to the water, energy and carbon cycles.

Many more satellite missions are in an advanced stage of planning, for example the joint India-France Megha-Tropiques Mission, and when launched they will provide a continuation of space-based measurements that are now being made and provide data on many new parameters.

Satellites cannot replace surface-based meteorological observations but can only complement them. There are several important parameters, like surface pressure for instance, which cannot be measured by satellites, and accurate in situ measurements will continue to remain very important for calibration and validation of satellite-derived products.

On the other hand, climate change related parameters can only be indirectly inferred from space-based measurements. Satellites are also very expensive to build. So the space-based systems for climate monitoring are essentially a compromise between conflicting factors like demands of science, technological feasibility, and availability of resources.

The record lengths of the satellite-derived data are inevitably much shorter than conventional data. The building up of long-term climate data based from space-based observations has therefore to be done with great care. This involves strict quality control of satellite data and derived products. Corrections have to be applied for degradation of on-board sensors with time and proper calibrations and inter-satellite comparisons have to be made when data sets obtained from different satellites over different time periods are to be merged into a single climate data series.

The International Satellite Cloud Climatology Project (ISCCP) was established as part of the World Climate Research Programme (WCRP) in 1982 with the prime objective is the production of a global, calibrated, and normalized data set of visible and infrared radiances along with basic information on the radiative properties of the atmosphere. The ISCCP initiative of 1982 was closely followed by the launch of another major international undertaking that complements the ISCCP. The Global Precipitation Climatology Project (GPCP) was established by the World Climate Research Programme (WCRP) in 1986 to address the problem of quantifying the long-term distribution of global precipitation change in tropical rainfall. Both these projects are examples of successful international cooperation towards compilation of extensive global data sets derived from satellites for use in climate research and climate change monitoring.

2.8 Paleoclimatology

To determine whether the current global warming trend is something unusual, it is essential to go as backwards in time as possible on the geological time scale and reconstruct the past climates that might have existed on the earth (Kelkar 2006). Reliable instrumental records are mostly available since the year 1850 or later. India has had a few stations like Chennai where observations have been recorded for the last 200 years or so. Estimates of global climate changes during the past can, therefore, be drawn only as indirect inferences from what are called proxy indicators.

A proxy indicator is a local record left by nature itself that can be interpreted now using physical or biophysical principles in terms of past climate events. It is necessary here to filter out the noise from the signal and eliminate false signals that may not have been related to climate. It is also desirable to match two or more independent proxy indicators for corroborating the inferences. Examples of proxy indicators are tree rings, corals, lake and ocean bottom sediments and bore hole measurements.

The most commonly found proxy indicators of past climates are tree rings, particularly in areas which are marked by seasonal changes in temperature and precipitation. The technique of assigning individual tree rings to past time periods and interpreting them in terms of the climate that might have prevailed then, has developed into a science of its own called dendrochronology. Suitable forest locations have first to be identified, then suitable trees have to be selected for slicing, and then the characteristics of the rings like width and density examined. Usually, data is collected from a large number of trees in one forest, which is then matched and combined. Tree ring analysis has yielded valuable information that has helped reconstruct past or paleoclimates. However, the reconstruction is obviously limited to the life of the trees sampled and cannot go back more than 1,000-2,000.years.

In one of the very early dendrochronology studies carried out in India, Pant (1979) had found that trees of the Pinus family such as Chir and other conifers along the Himalayan snowline and sub-Himalayan mountains had a prominent ring structure and they showed a significant response to temperature and snowfall. A lot of work has since been done by Indian scientists on tree rings in the Western Himalayas and other parts of India, particularly with regard to simulating the behaviour of the monsoons in the past.

Yadav (2009) have confirmed that tree ring analyses from semi-arid and arid regions in western Himalaya have an immense potential for reconstructing past temperatures on a millennial time scale. However, paucity of weather records from stations close to tree ring sampling sites poses difficulty in calibrating tree ring data against precipitation because of its variability. Tree ring records from the region indicate multi-century warm and cool anomalies consistent with the Medieval Warm Period and Little Ice Age conditions.

Indian scientists have also tried to analyse tree rings in central and peninsular Indian forests. Borgaonkar et al (2009) have compiled a 523-year tree ring width index chronology of teak trees dating back to 1481 from three forest sites over Kerala in southern India, Dendroclimatological investigations indicate a significant positive relationship between the tree ring index series and Indian summer monsoon rainfall and related global parameters like the southern oscillation index. A higher frequency of occurrence of low tree growth is observed in years of deficient Indian monsoon rainfall and droughts associated with El Nino years since the late 18th century. Prior to that time, many low tree growth years are detected during known El Nino events, probably related to deficient Indian monsoon rainfall. The Kerala tree ring chronology have a high degree of sensitivity to monsoon climate and is a useful tool to extend our understanding of the vagaries of monsoon rainfall over periods for which historical meteorological records are not available.

Ram et al (2009) have developed tree ring chronologies for the 19th and 20th centuries for teak from three different tree sites in Madhya Pradesh and Andhra Pradesh in Central India. They found a significant positive correlation of the tree ring width with an April-September moisture index.

Corals found in the shallow waters of tropical and subtropical oceans are another source of information that can complement tree ring data that can obtained only over land. Like trees, corals too have a ring structure, and a new ring gets added every year. The ring structure can be analysed by drilling into the coral's core. Other proxy data are those derived from lake and ocean floor sediments, at latitudes poleward of the tree line, annually laminated lake bed sediments, and such other high resolution data including bore hole measurements.

A storehouse of information about the past climates of the earth lies hidden under the ice in the polar regions of Canada, Greenland, and Antarctica and other alpine and subtropical regions. This information can be unraveled by drilling deep into the ice and extracting ice cores which can be later subjected to an analysis of oxygen isotopes, concentration of salts and accumulation of precipitation, both annual and seasonal.

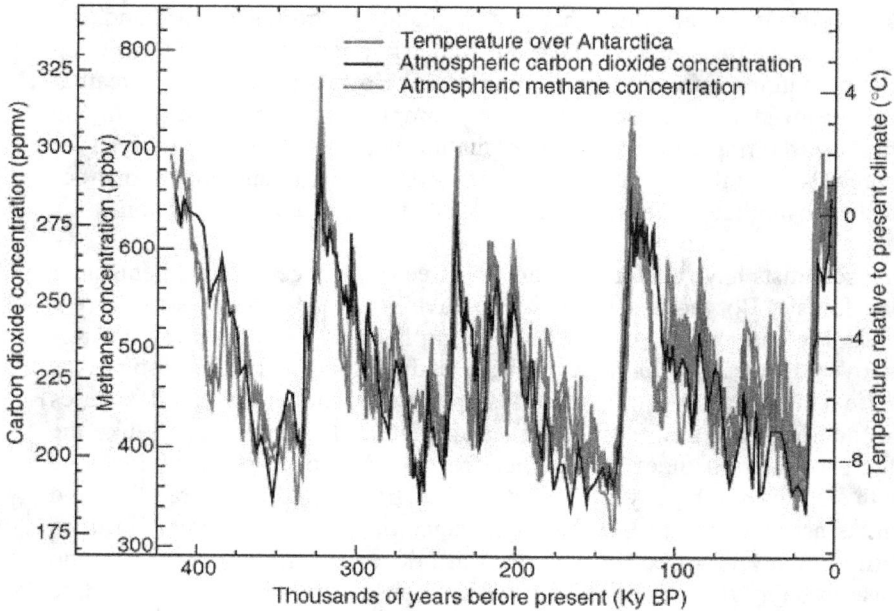

Figure 2.8.1 Variations of temperature, and methane and atmospheric carbon dioxide concentrations derived from air trapped within ice cores from Antarctica (Source: IPCC TAR, Folland et al 2001)

Results of very great significance have emerged from the analysis of the 420,000-year records of the ice cores recovered at the Vostok station in eastern Antarctica by Petit et al (1999) who have extended the ice record of atmospheric composition and climate to the past four glacial–interglacial cycles. The succession of changes through each climate cycle and termination was similar, and atmospheric and climate properties oscillated between stable bounds. Interglacial periods differed in temporal evolution and duration. Atmospheric concentrations of carbon dioxide and methane correlate well with Antarctic air temperature throughout the record. Present day atmospheric burdens of these two important greenhouse gases seem to have been unprecedented during the past 420,000 years (Figure 2.8.1).

The temporal dynamics of global temperature and of the global carbon cycle, as represented by the atmospheric concentration of CO_2 and methane are tightly coupled and show very similar patterns of variation throughout the period of the record. Four cycles of 100,000-year periodicity are clearly seen. The values fall recurrently within the same envelope through the four cycles over the last half a million years. This systemic behaviour of the earth's environment is due to a combination of external forcing, primarily variations in solar radiation levels at the earth's surface, internal forcings within the earth's environment itself and a large and complex array of feedbacks. An important outcome of the ice core data analysis is the realization that it is the internal dynamics of the system, rather than external forcings (IPCC TAR 2001, Folland et al 2001).

2.9 References

Borgaonkar H. P., Sikder A. B., Somaru Ram, and Pant G. B., 2009, "El Nino and related monsoon drought signals in 523-year-long ring width records of teak (Tectona grandis L. F.) trees from south India", *Palaeogeography, Palaeoclimatology, Palaeoecology*, doi:10.1016/ j.palaeo.2009.10.026.

Brohan P., Kennedy J. J., Harris I., Tett S. F. B. and Jones P. D. 2006, "Brohan Uncertainty estimates in regional and global observed temperature changes: a new dataset from 1850", *J. Geophysical Research,* 111, D12106.

Folland C. K. and coauthors, 2001, "Observed climate variability and change", *Climate Change 2001: The Scientific Basis. Contribution of Working Group I to the Third Assessment Report of the Intergovernmental Panel on Climate Change* [Ed: Houghton J. T. et al], Cambridge University Press, 99-182.

Hansen J. and coauthors, 1981, "Climate impact of increasing atmospheric carbon dioxide." *Science*, 213, 957-966.

Hansen J. E. and coauthors. 2001, "A closer look at United States and global surface temperature change". *J. Geophysical Research*, 106, 23947-23963.

Kelkar R. R., 2006, "The Indian monsoon as a component of the climate system during the Holocene", *J. Geological Society of India*, 68, 347-352.

Komhyr W. D. and coauthors, 1989, "Atmospheric carbon dioxide at Mauna Loa Observatory: 1. NOAA GMCC measurements with a non-dispersive infrared analyzer" , *Journal Geophysical Research*, 94, 8533-8547.

Pant G. B., 1979, "Role of tree-ring analysis and related studies in palaeo-climatology: preliminary survey and scope for Indian region", *Mausam*, 30, 439-448.

Petit J. R. and coauthors, 1999, "Climate and atmospheric history of the past 420,000 years from the Vostok ice core, Antarctica", *Nature*, 399, 429-436.

Smith T. M. and Reynolds R. W., 2005, "A global merged land air and sea surface temperature reconstruction based on historical observations (1880-1997)", *J. Climate*,18, 2021-2036.

Sonbawne S. M., Ernest Raj P., Devara P. C. S. and Dani K. K., 2009, "Variability in sun photometer derived summertime total column ozone over the Indian station Maitri in the Antarctic region", *International Journal of Remote Sensing*, 30, 4331-4341.

Yadav R. R., 2009, "Tree ring imprints of long-term changes in climate in western Himalaya, India", *Journal of Biosciences,* 34, 699-707.

Zhao C. and Tans P., 2006, "Estimating uncertainty of the WMO mole fraction scale for carbon dioxide in air", *Journal Geophysical Research*, 111, D08S09, doi: 10.1029/2005JD006003.

Chapter 3

Climate Modelling and Prediction

In any discipline or profession, models are a link between conception and perception, between ideas and reality, between plans and implementation. Architects make smaller replicas of the structures they design for people to see. Scientists organize laboratory experiments to prove their hypotheses and theories. Engineers make prototypes before making the actual machines. Fashion designers use people as models to show their creations.

Models help us to see what is otherwise difficult to visualize and that is the reason why the use of numerical models has come in a big way into atmospheric and ocean sciences. A few decades back laboratory experiments were attempted for simulating atmospheric motions in a dishpan and to create clouds in a chamber. Geophysical fluid dynamics and cloud physics have made excellent progress since then, but the greatest limitation to replicating clouds and weather in a laboratory is imposed by their size and variability. Scientists had long ago known the basic hydrodynamical laws that govern the motions of the atmosphere and oceans, but they could not apply them in day to day weather forecasting because of the tedious computations required for solving the equations. The time taken to generate such predictions was so long that the effort would be rendered futile.

The real breakthrough in atmospheric modelling came in the 1980s with the availability of supercomputers and was soon followed by developments in ocean modelling. With the exponential increase in computing power and the speed of global data collection, atmosphere and ocean models have since been being growing in their complexity, reliability and scope.

What numerical weather prediction models can do in short range weather prediction could perhaps be done equally well by highly skilled synoptic meteorologists, but for medium range and extended range predictions, there is no alternative to using numerical models. And when it comes to climate, only models can help us to understand its intricacies and make us capable of predicting the climate of the future.

One of the biggest advantages of using models is that they allow us to perform experiments that would be impossible to conduct in the real atmosphere or ocean, In atmospheric and ocean model, physical processes are represented entirely by a set of equations, the state of atmosphere and ocean is prescribed as a set of initial conditions and the model solves the equations to get the future state. If we run the model with different initial or boundary conditions or modify the equations, and compare the model results, we can get an idea of how the atmosphere or ocean will react in reality to such forcings.

One of the earliest numerical simulations of the monsoon was made by Murakami, Godbole and Kelkar (1970) at the Institute of Tropical Meteorology, Pune. They simulated a near-steady state monsoon in a zonally symmetric framework along the longitude 80° E in the month of July. They designed a primitive equation model with a σ-system in which the atmosphere was divided into 8 layers. The top of the frictional boundary layer was taken as level 8½. In the horizontal only y-variation was considered, with 18 grid points at 5° latitude interval. In the x-direction, symmetry was assumed. Radiation heating and cooling was computed within the model according to the scheme described by Godbole, Kelkar and Murakami (1970), using the distribution of absorbing gases and clouds for the month of July. During the iterations, whenever the vertical lapse rate of temperature exceeded the wet (dry) adiabatic value with saturated conditions, it was adjusted to the wet (dry) adiabatic value by a redistribution of static energy. SST was kept at a constant value of 300 °K, while the land surface temperature was computed from the heat balance of the surface fluxes. The model was run with a calm initial state and the initial vertical distribution of temperature was prescribed as per the standard atmosphere. The model was run with a 10-minute time step for 80 days and experiments were repeated without the Himalayan orography.

Murakami, Godbole and Kelkar (1970) were successful in simulating some of the large-scale features of the monsoon circulation and the wind speeds were of the same order as those in climatological normals. Their result of particular importance was that when the Himalayan mountains were removed from the orographic profile along the longitude 80° E, the zonal circulation became much weaker, with lower level westerly winds of 10 knots and upper level easterly winds of 20 knots only. Thus a realistic parameterization of the Himalayan mountains is crucial to the numerical simulation of the monsoon.

Subsequent to the pioneering work of Murakami et al, there have been several modelling studies of the monsoon that have been able to simulate reasonably well many features of the monsoon such as the tropical easterly jet, the Somali jet and the low level westerly flow over India. However, even as of today, the most challenging problem is that of simulating realistically the mean rainfall pattern of the monsoon.

3.1 Types of Climate Models

For gaining an insight into the processes that operate in the climate system of the earth and for using that knowledge for predicting the climate of the future, climate models are indispensable. A variety of climate models are currently available and it is important to choose a model judiciously so as to serve the purpose and scope of the investigation.

When dynamical models are used for making weather predictions several days ahead, the accurate specification of the initial conditions of the atmosphere is a prime necessity. Climate models, however, do not take into consideration the evolution of weather systems on such a short time scale and the initial conditions become relatively unimportant. On the other hand, The accuracy of climate predictions depends more on how well the parameterization of radiative forcings and ocean processes is done in the model, Climate models have been steadily improving in these respects and some models treat aerosol-related radiative forcing interactively.

The dynamical models that are currently in use differ widely with regard to the complexity of their structure and the manner in which they incorporate the physical processes of the atmosphere-land-ocean system (Meehl et al 2007, Solomon et al 2007). The Simple Climate Models (SCMs) are built around an energy balance equation, a prescribed value of climate sensitivity and a basic representation of ocean heat uptake. Many of the simple models are of the slab ocean type in which the dynamics of the ocean are not considered at all. However, SCMs are still useful because they can provide a broad estimate of the increase in global mean temperature and the rise in sea level due to thermal expansion. SCMs are useful in conceptual inter-disciplinary exercises as they can be further coupled to other simplified models to study land-atmosphere, climate-biosphere and such other interactions. It would be very difficult to construct models that have full details of all the processes in such diverse domains. That is the reason why even with the increasing availability of computing power, SCMs are still in vogue.

Compared to SCMs, the Earth System Models of Intermediate Complexity (EMICs) have better parameterizations of the atmospheric and oceanic circulations and biogeochemical cycles. The possible long-term changes in climate can be estimated by running the EMIC models over several centuries in an ensemble mode. The models can even be run on time scales that are representative of interglacial oscillations, but such runs are computationally very expensive and a compromise has to be made in terms of the horizontal resolution. EMICs are preferred over the more complex models for making climate sensitivity experiments and for generating probabilistic climate projections over a wide range of emission scenarios (Randall et al 2007).

At the higher levels of complexity are the Atmospheric General Circulation Models (AGCMs), the Ocean Circulation Models (OCMs) and the fully coupled Atmosphere-Ocean General Circulation Models (AOGCMs). AOGCMs incorporate the dynamics of the atmosphere and oceans as well as land surface processes, and they include sea ice and other components. AOGCMs are currently being run by several centres around the world and over 20 AOGCMs have been used in organized climate-related experiments.

One of the difficulties in running atmosphere-ocean coupled models is that the initial state of the ocean is known much less precisely compared to that of the atmosphere. Another problem is their tendency to drift away from a realistic state because of small flux imbalances which need to be corrected. Most of the contemporary models are stable enough without such flux adjustments.

3.2 Problems with Climate Models

While climate projections made by different models tend to agree qualitatively, they most often differ quantitatively in many respects. Although the dynamics of AOGCMs are becoming increasingly comprehensive, the uncertainties in the parameterization of physical processes are the principal cause of the differences between model results. Another cause is that the lower resolutions of some of the models do not allow the finer scale features to be captured by them. Therefore, when it comes to taking practical action, it becomes imperative to judge how much the model results are to be believed, which results could be assigned a higher credibility than others, and what choice to make when the results are conflicting. For this purpose, many international efforts have been launched to compare the performance of different models in experiments performed under common conditions. Since the Atmosphere-Ocean General Circulation Models (AOGCMs) are being used as the primary tool for simulating past climates and making future climate projections, it is important to apply norms for evaluating the models and interpreting their results (Solomon et al 2007).

A crucial issue in the acceptance of climate model projections is to know how realistically the model is capable of simulating the mean climate as we presently know it. If a climate model cannot pass this benchmark test, it goes down in the level of credibility that can be assigned to its future projections. Here again, there is no single climate model that can faithfully simulate the present climate in every respect. For example, some models may be good in simulating the monsoon rainfall pattern on a spatial or temporal scale in a relative manner but not the rainfall amounts in absolute terms. Then there could be some models that correctly simulate the monsoon seasonal rainfall but fail in the distribution over each of the four months. It is necessary to evaluate the strengths and weaknesses of various models with regard to such details which may not be important on a global scale, but could be crucial on a regional or local scale.

In recent years, model simulations of the present climate have undoubtedly improved in general, but compared to sea level pressure and surface temperature, deficiencies still remain in the simulation of tropical precipitation. As regards the simulation of climate variability, most climate models are now capable of simulating the dominant modes of extra-tropical climate variability but many models still have problems simulating ENSO.

While interpreting the results of any climate model, it is of vital importance to know what type of model it is, its limitations, past history, and its strengths and weaknesses. Crude models would obviously produce crude results. Sophisticated and advanced models might bring out conflicting features in their prediction, of which some may seem acceptable but not all. Particularly when sensational and scary predictions are made, like say the metropolis of Mumbai would get submerged by 2100, or the Indian monsoon would dry up by 2060, the models on which they are based must themselves be thoroughly investigated to ascertain their credibility.

Even on the seasonal and subseasonal scales of the monsoon rainfall, there are some models which perform better in terms of the rainfall pattern, while some other models may perform better in terms of the rainfall amounts. There may be models which are good at predicting only droughts. Although the merits and demerits of the models can be compared in hindsight, the problem arises as to which model to believe out of the several alternatives when it comes to real time prediction. Diagnostic studies are no doubt important but they do not help much for purposes of prognostication.

3.3 Climate Predictions and Projections

Climate models that parameterize the physical processes of the atmosphere, land and ocean and the interactions between them, can be used to predict the future state of the climate system assuming that there are no human influences on climate. This assumption is by itself invalid, because it is contrary to the basic premise that the current global warming is anthropogenic. The role of human society in modifying the future climate of the earth is extremely difficult to envisage. Climate models that are based upon purely physical processes themselves have several limitations and uncertainties. Now to expand the scope of these physical models to include the thought processes of the human mind would be a stupendous task. However, this is exactly what needs to be done in order to arrive at an accurate prediction of the state of the earth's climate say a hundred years from now. An ideal climate model must have the ability to quantify the interactions between human beings and the physical earth, to foresee the evolution of human society over the next century or two, and to envision how nations will cope with the problem of climate change or whether they will allow it to aggravate.

As of now it is seemingly impossible to construct such elaborate models, but we can come close to finding a practical and easier solution to the problem by resorting to the use of what have come to be known as emission scenarios. The human mind is extremely innovative, the human body can learn to adapt to nature and human beings have the capability to surmount natural obstacles coming in the way of progress. Therefore, there is no doubt that human ingenuity will eventually prevail over global warming. How and when this will happen is an open question, but broad global scenarios can be envisaged in terms of the growth of the world's population, the pattern of energy consumption, the release of greenhouse gases into the atmosphere, technological progress, and such other factors. Since the predictions will vary depending upon what scenario is chosen, it has become a practice to call them climate 'projections' to distinguish them from climate 'predictions' made on purely physical considerations.

3.4 Emission Scenarios

As the greenhouse gas concentrations in the atmosphere are at least partly a measure of the influence of human actions on climate, the scenarios used for climate projections are termed as 'emission scenarios'. Future GHG emissions can be visualized as an outcome of driving forces of different types, demographic, socio-economic or technological. Some of these driving forces can be quantified and their future magnitudes can be extrapolated using appropriate techniques. Theoretically speaking, an infinite number of emission scenarios can be constructed, but by ruling out the very unlikely ones, a finite number of scenarios can be chosen. As aptly stated by Grubler et al (2001) each future scenario is path-dependent and results from a large series of conditionalities. The scenarios must also be internally consistent.

Socioeconomic variables cannot be randomly combined or interchanged as they are also interdependent.

In 1992. the IPCC released emission scenarios to be used by climate models. The IPCC decided in 1996 to develop a new set of emissions scenarios for its Third assessment Report and a Special Report on Emissions Scenarios (SRES) was published (Nakicenovic et al 2000). These IPCC scenarios have been used for running climate models and they have come to be known as the 'SRES scenarios'. They can serve as a basis for climate prediction purposes, as well as for climate change analysis and the assessment of impacts, adaptation, and mitigation.

The IPCC SRES scenarios have four different narrative 'storylines' which were developed to describe consistently the relationships between emission driving forces and their evolution. Each of the four storylines represents different demographic, social, economic, technological, and environmental developments, both positive and negative. Within a storyline, there are individual scenarios which represent specific quantitative interpretations of the storyline. Individual scenarios that follow the same storyline can be grouped together to form a scenario 'family'. It is important to note here that the IPCC SRES scenarios do not include climate mitigation initiatives, envisaged under the UNFCCC and the Kyoto Protocol.

In all, there are 40 SRES scenarios which form four families named as A1, A2, B1 and B2.

The A1 family of emission scenarios refers to a more integrated world characterized by rapid economic growth, a global population that reaches 9 billion in 2050 and then gradually declines, and the quick spread of new and efficient technologies. The A1FI scenario visualizes a high consumption of fossil fuels like coal, oil and gas. The A1T scenario relies predominantly on non-fossil fuel energy sources. The A1B scenario strikes a balance amongst all energy sources, not relying too heavily on one particular energy source, on the assumption that similar improvement rates apply to all energy supply and end use technologies.

The A2 family of scenarios represents a heterogeneous and more divided world consisting of independently operating, self-reliant nations that preserve their local identities. There is a continuous growth in the population, regionally oriented economic development, slower and more fragmented technological changes and increase in the per capita income.

The B1 scenarios are of a more integrated, eco-friendly and stable world having the same population and economic growth as in A1, but with rapid changes towards a service and information economy, and introduction of clean and resource efficient technologies.

The B2 scenarios visualize a world that is more divided, but more eco-friendly, population increasing at a slower rate than in A2 and intermediate levels of economic and technological development.

A1FI, A1B, A1T, A2, B1 and B2 are the six most commonly used emission scenarios in the context of future climate change. The total global annual CO_2 emissions from all sources including energy, industry and land use change, from 1990 to 2100 associated with these six scenarios is shown in Figure 3.4.1.

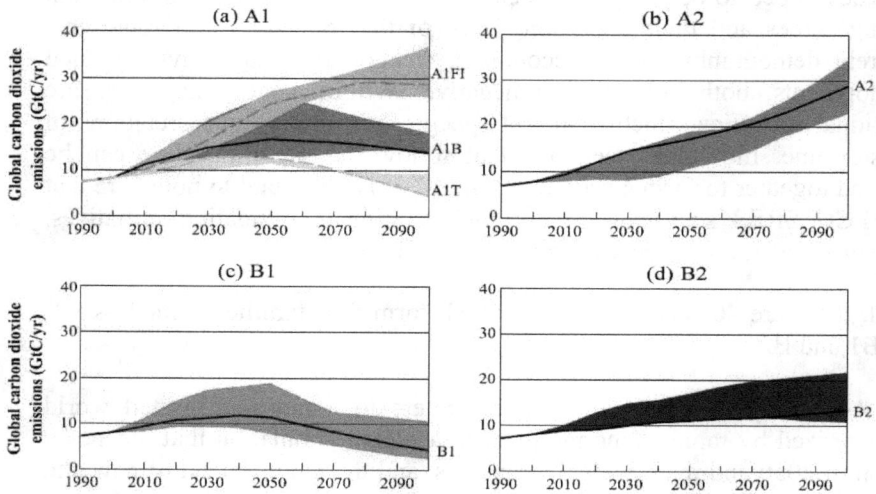

Figure 3.4.1 Total global annual CO_2 emissions from all sources from 1990 to 2100 (in Gt Carbon/yr) for the four scenario families. The coloured bands show the range of scenarios within them and the black line is an illustrative scenario. (Source: IPCC 2000)

3.5 IPCC Climate Projections for the 21st Century

The climate projections made by the IPCC (IPCC AR4, Meehl et al 2007) are derived from a hierarchy of model runs using the various emissions scenarios. The general projection is that global warming and induced climate changes during the 21st century would very likely be larger than those observed during the 20th century. However, in quantitative terms there is a large uncertainty in this result because of the differences in the emission scenarios. There is no guarantee that any one of the 40 emissions scenarios will be realized in the actual world.

The projected global surface temperature rise during the period 2090-2099 compared to 1980-1999 for six commonly used SRES scenarios and with the GHG concentrations held constant at their levels in the year 2000 are given in Table 3.5.1. Figure 3.5.1 shows the likely evolution of the warming over time. The best estimates and corresponding likely ranges have been derived from a hierarchy of climate models, ranging from simple to complex AOGCMs and using lower to higher SRES emission scenarios. The temperature rise is likely to be 1.1 to 2.9 °C for the low emission B1 scenario and 2.4 to 6.4°C for the high emission A1FI scenario.

While Table 3.5.1 and Figure 3.5.1 give the average values for the globe as a whole, Figure 3.5.2 shows the global distribution of the projected surface temperature changes for the early and late 21st century relative to the period 1980-1999. These again are multi-model means using the moderate A1B emissions scenario. A significant feature of the figure is that the projected 21st century temperature change is positive everywhere. It is greatest over land and at most high latitudes in the northern hemisphere during winter, and increases from the coasts towards the interior regions of the continents. In geographically similar areas, warming is typically larger in arid than in moist regions. In contrast, warming is least over the southern oceans and parts of the North Atlantic Ocean.

What is most evident from Figures 3.5.2 and 3.5.3 is that the entire Asian monsoon region is likely to experience a warming accompanied by an increase in precipitation rate in the June-July-August season. Over many parts of India, especially the western peninsula and northeast India, the monsoon precipitation rate is likely to register an increase of the order of 10-20 % over the current rate by the end of the century.

**Table 3.5.1 Global Surface Temperature Change (°C)
during 2090-2099 relative to 1980-1999 for Various Emissions Scenarios
(Source: IPCC AR4, Meehl et al 2007)**

SRES Scenario	Best Estimate (°C)	Likely Range (°C)
B1	1.8	1.1-2.9
A1T	2.4	1.4-3.8
B2	2.4	1.4-3.8
A1B	2.8	1.7-4.4
A2	3.4	2.0-5.4
A1FI	4.0	2.4-6.4

Global surface warming (°C)

6.0
5.0
4.0
3.0
2.0
1.0
0.0
−1.0

A2
A1B
B1
Year 2000 Constant
Concentrations
20th century

B1 A1T B2 A1B A2 A1FI

1900   2000   2100

Year

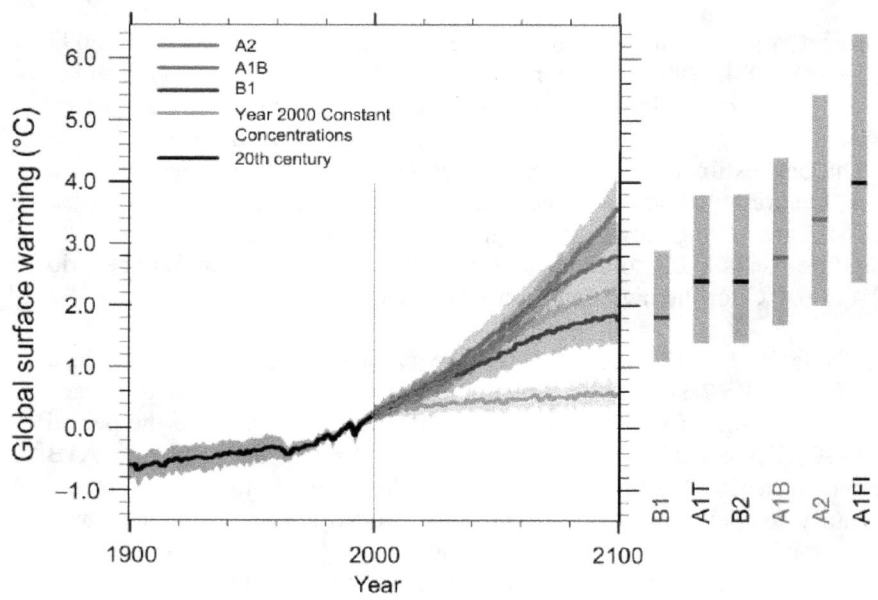

Figure 3.5.1 Multi-model global annual average surface temperature anomaly relative to 1980-99 for different emission scenarios. Shading denotes ± 1 standard deviation range. The orange line is the 21st century projection with GHG concentrations held constant at year 2000 values. The gray bars on the right indicate the uncertainty range. (Source: IPCC AR4 2007)

What is significant is that since these results are obtained keeping the GHG concentrations frozen at the 2000 levels, the earth is not going to cool down until 2100 even if stop adding GHGs to the atmosphere from now onwards, which is also a hypothetical case. So the real remedy lies only in reducing the GHG concentrations in order to bring the temperature down. In fact, models show that irrespective of the emission scenario chosen, the earth is likely to keep warming at a rate of about 0.1-0.2 °C per decade for the next two decades, which can be attributed mainly to the slow response of the oceans.

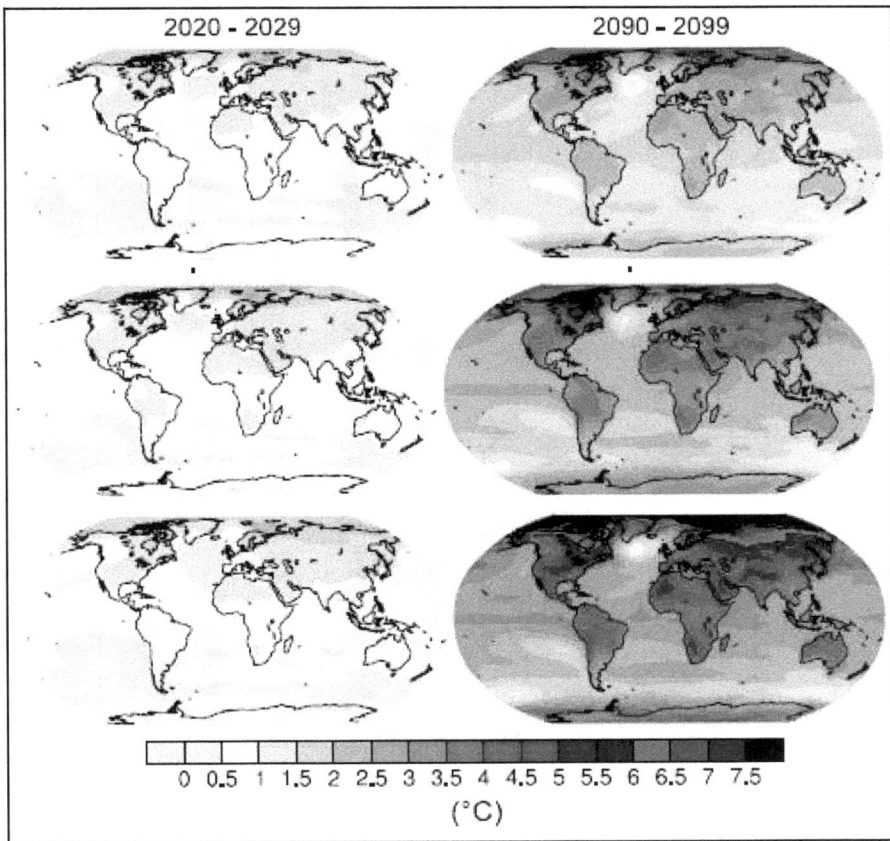

Figure 3.5.2 AOGCM multi-model average projections of the global distribution of surface temperature rise (°C) for the B1 (top), A1B (middle) and A2 (bottom) SRES scenarios averaged over the decades 2020 to 2029 (left) and 2090 to 2099 (right) relative to the period 1980-1999 (Source: IPCC AR4, Solomon et al 2007)

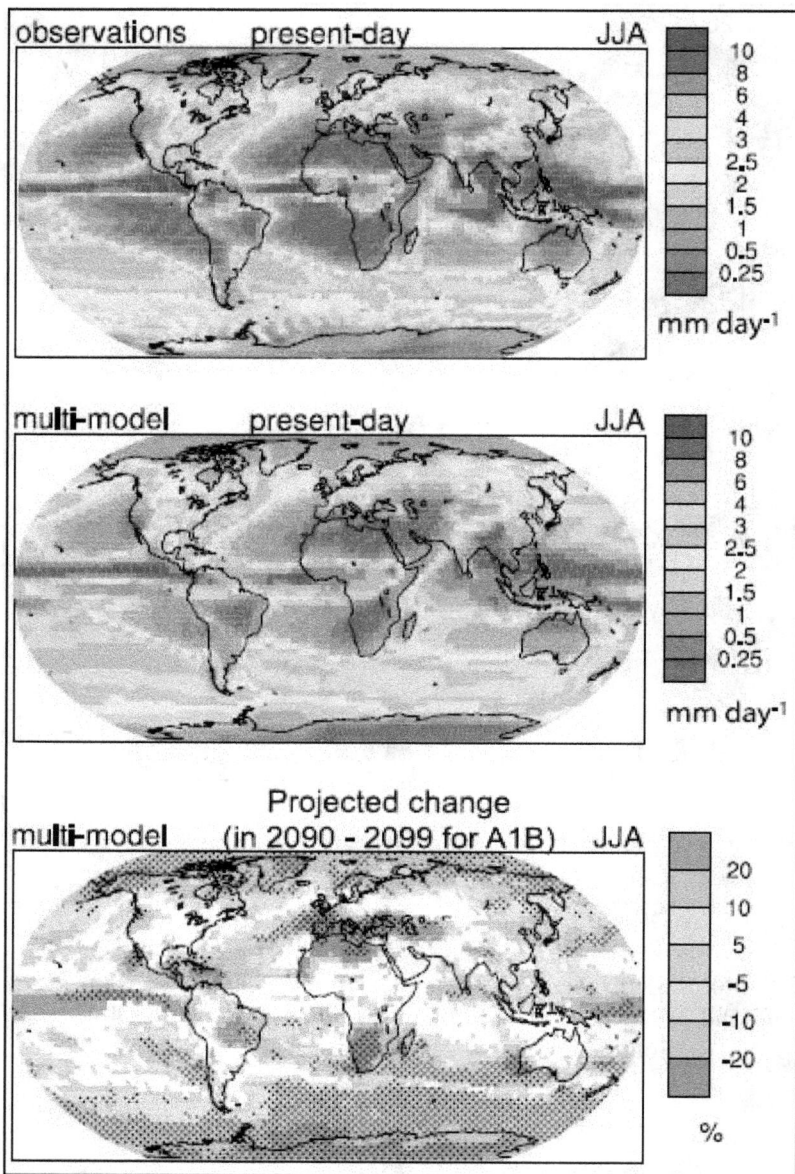

Figure 3.5.3 Global distribution of June-July-August mean precipitation rate in mm/day for the period 1979-1993: observed (top) and MMD A1B simulation (middle) and projected change in % (bottom) for the period 2090-2099 relative to 1980-1999. (Source: Solomon et al 2007)

This raises the question as to how much addition of GHGs can be allowed into the earth's atmosphere if we were to set a limit of say not more than 2 °C to a further rise in the temperature. Many countries of the world have adopted such a limit of 2 °C temperature rise over the pre-industrial level for planning their own mitigation actions. Several components of the carbon cycle are not yet completely understood and the response of the climate system to changes in the carbon cycle is also not known with certainty. Hence the issue cannot be resolved easily and the answers will be model-dependent.

In one such study (Meinshausen 2006) have concluded from a set of 11 published climate sensitivity uncertainty estimates, that if the CO_2 concentration is allowed to stabilize at 550 ppm, there is a 63-99 % probability of the temperature exceeding 2 °C. Only if the CO_2 level peaks at 400 ppm, the equilibrium temperature rise could possibly be contained within 2 °C. Such results often become issues for discussion at climate change negotiations, and the CO_2 peaking value is currently a matter of debate.

In a further study, Meinshausen et al (2009) have shown that limiting the cumulative CO_2 emissions over 2000-50 to 1,000 Gt CO_2 yields a 25 % probability of warming exceeding 2 °C and a limit of 1,440 Gt CO_2 yields a 50 % probability. The probability of exceeding 2 °C goes up to 53-87 % if global GHG emissions in 2020 remain 25 % above their 2000 levels.

3.6 References

Godbole R. V., Kelkar R. R. and Murakami T., 1970, "Radiative equilibrium temperature of the atmosphere along 80° E longitude", *Indian J. Meteorology Geophysics*, 21, 43-52.

Grübler A. and Nakicenovic N., 2001, "Identifying dangers in an uncertain climate", *Nature*, 412, 15.

IPCC, 2000, *Summary for Policymakers, Working Group III Special Report on Emissions Scenarios*, Intergovernmental Panel on Climate Change, 27 pp.

IPCC, 2007, *Summary for Policymakers, Working Group I Report to the Third Assessment Report,* Intergovernmental Panel on Climate Change, 18 pp.

Meehl G. A. and coauthors, 2007, "Global Climate Projections". *Climate Change 2007: The Physical Science Basis. Contribution of Working Group I to the Fourth Assessment Report of the Intergovernmental Panel on Climate Change* [Ed: Solomon S. et al], Cambridge University Press, 748-845.

Meinshausen M., 2006, "A brief analysis based on multi-gas emission pathways and several climate sensitivity uncertainty estimates", Avoiding Dangerous Climate Change, 265-280.

Meinshausen M. and coauthors, 2009, "Greenhouse-gas emission targets for limiting global warming to 2 °C", *Nature*, 458, 1158-1162.

Murakami T., Godbole R. V. and Kelkar R. R., 1970, "Numerical simulation of the monsoon along 80° E", *Proc. Conf. Summer Monsoon of Southeast Asia*, Norfolk, Va, USA, [Ed: Ramage C. S.]. 39-51.

Nakicenovic N. and Swart R., 2000 (Ed.), *Special Report on Emissions Scenarios*, Cambridge University Press, 612 pp.

Randall D. A. and coauthors, 2007, "Climate models and their evaluation", *Climate Change 2007: The Physical Science Basis. Contribution of Working Group I to the Fourth Assessment Report of the Intergovernmental Panel on Climate Change* [Ed: Solomon S. et al], Cambridge University Press, 590-662.

Solomon S. and coauthors (Ed), 2007, *Technical Summary, Climate Change 2007: The Physical Science Basis. Contribution of Working Group I to the Fourth Assessment Report of the Intergovernmental Panel on Climate Change*, Cambridge University Press, 91 pp.

Chapter 4

Climate Change Impacts on Glaciers, Sea Level and Tropical Cyclones

If climate predictions are to be acted upon, which is indeed their purpose, those who make them must provide the users with an honest self-assessment of the likelihood of their predictions proving correct. More often than not, this is not done, particularly so when the results of scientific investigations of climate change get reported through the media. The scientific studies are converted into stories and complex issues are oversimplified. Viewers or readers could easily get convinced that what is presented in the media stories is the real truth and what is foretold in the name of climate change is sure to happen. They are not in a position to raise questions nor do they have the means of cross checking on the reliability of the source of information.

A similar situation may arise even at a higher governmental and political level, when a country's long term policies are being formulated or critical decisions are required to be taken in which climate change comes into picture. It may happen that during the decision-making process, the climate predictions that are available for reference come from sources outside their lands who have little knowledge of the country's specific situation and whose credibility is unknown or unproven.

Further, there is a problem of scale which often gets disregarded in the interpretation of climate change predictions. Very commonly global scale results or predictions derived using models that have very low spatial resolutions of the order of 500 or 1,000 km, are applied directly to the regional or local scale, and incorrect and unrealistic conclusions are drawn. On the contrary, some local extreme events are linked and extrapolated and made to appear as if they are occurring universally, which is also wrong.

Finally, the results of several climate change investigations are qualitatively meaningful and logically acceptable, like warming will cause ice to melt and the sea level to rise, but when it comes to quantifying the processes, they may have large error bars. It is important to know the error bars especially

73

when the quantities are small, while considering the practical application of the scientific results in real situations.

4.1 India's Major Concerns

India is a vast country with a population of over a billion and a variety of climates and topographical features. It depends on the monsoons for meeting all its water needs, with its major rivers either originating from the Himalayas or fed by the monsoon rains. In the past it has had to face severe droughts, widespread floods and devastating tropical cyclones. Sixty per cent of the people are directly dependent upon agriculture for their livelihood. Life in India is, in some way or the other, linked to the climate, and the threat of climate change therefore looms large over India in a real sense. Although the problem of climate change has global dimensions, there are certain challenges that are specific to India and with which the rest of the world may not be much concerned. India will also like to see that the mitigation actions required to be taken to lessen the severity of the impacts do not become a stumbling block in the path of sustainable development.

Therefore, India has reason to be concerned about the likely impacts of climate change on at least six major fronts. These are:

1) Retreat of Himalayan glaciers and its effect on the Indian rivers which originate in the Himalayas

2) Threat of sea level rise to its 7,500 km long coastline and the Andaman, Nicobar and Lakshadweep islands

3) Possible increase in the frequency and intensity of tropical cyclones over the Bay of Bengal and Arabian Sea, which cause heavy losses of life and property in the coastal regions

4) Change in the pattern and quantum of monsoon rainfall, which is India's only source of water

5) Effect on temperature rise and change in the rainfall pattern on agricultural production and its impact on food security

6) The possible effect on the health of the population arising from the growth of vector-borne diseases due to rising temperatures

The first three issues will be discussed from both global and Indian perspectives in this chapter and the other three will be addressed in the next chapter.

There are several other areas in which climate change may have an impact, such as forest cover, biodiversity, natural ecosystems, climate-sensitive industries, population migration, and more. These are beyond the scope of this book.

4.2 Concepts and Terminology

In the process of evolution of the older and simpler climatology into the present complex science of climate change, a whole new vocabulary has come into being. Therefore, to begin with, it would be desirable to get acquainted with some of the terminologies and concepts that are commonly encountered in the discussions about climate change impacts.

Impacts of climate change: These are the possible consequences of climate change on natural and human systems. They may not necessarily be harmful as they are often made out to be, but could also be beneficial. *Potential impacts* are those that are likely to occur under a projected climate change scenario. However, it is possible that some of the likely impacts could be partially offset because of the parallel process of natural and human adaptation. This will result in reduced impacts, which are referred to as *residual impacts*. For India, the sectors in which climate change impacts are most likely to be felt have been given above.

Vulnerability: Vulnerability is defined as the degree to which a system is susceptible to, or unable to cope with, the adverse effects of climate change, including climate variability and extremes. It is also a function of the character, magnitude and rate of climate change to which a system is exposed, and its sensitivity and adaptive capacity. In order to be prepared to face the effects of future climate change, it is prudent to make an assessment of India's vulnerability in various sectors like agriculture, energy, industry, and in different geographical zones.

Adaptation: Many times climate change is projected as a threat to nature or humanity, but both natural and human systems have an inherent capability to adjust to a new or changing environment, and such an adjustment is called adaptation. Since climate change operates on the scale of a century or decades, the parallel long-term process of natural and human adaptation will help to moderate the adverse effects of climate change and it may also be able to exploit beneficial opportunities.

Adaptive capacity: This is the degree to which natural and human systems are actually capable of adjusting to climate change, including climate variability and extremes.

Mitigation: This term is usually applied specifically to human intervention to reduce the sources of GHGs in the atmosphere and to enhance their sinks. However, it is often used with a more general and wider meaning to describe actions for reducing the harmful effects of climate change.

Likelihood and probability: The Intergovernmental Panel on Climate Change (IPCC) in its Fourth Assessment Report (AR4) has used a 10-point scale to denote various specific degrees of likelihood of a future event or the outcome of a process, in relation to the probability of its occurrence (Table 4.2.1).

**Table 4.2.1 Likelihood and Probability of Occurrence
(Source: IPCC AR4, Solomon et al 2007)**

Likelihood	Probability of occurrence
Virtually certain	Greater than 99%
Extremely likely	Greater than 95%
Very likely	Greater than 90%
Likely	Greater than 66%
More likely than not	Greater than 50%
About as likely as not	Between 33% and 66%
Unlikely	Less than 33%
Very unlikely	Less than 10%
Extremely unlikely	Less than 5%
Exceptionally unlikely	Less than 1%

The subtle nuances of the likelihood terminology need to be carefully understood. For example, when an event is said to be 'likely', it is implicit that there is a 1-to-3 chance of its non-occurrence, and this is far from negligible. It may also be prudent to go a step further and find out more about the method used by the investigators to arrive at the probability itself. Developing such awareness will be of great value when actions have to be initiated in the present time towards averting the potential consequences of the occurrence of a future event.

4.3 Uncertainties in Climate Change Science

The discussion in this book about the likely climate change impacts on India in the six crucial domains is based upon what is known about them today. Since we are considering a scenario that spans the next 80 to 100 years, there are so many imponderables, and the future is uncertain and unpredictable in so many respects. It is therefore impossible to fully visualize how nature will actually behave over the coming century, how human societies will evolve and what developments will take place in science and technology. Only certain scenarios can be assumed but they may or may not be realized. The impacts of climate change on India and the vulnerability of its population, and the mitigation strategies that we adopt, will all have to undergo a constant process of refinement as time goes by and course corrections made as we know better.

In fact, the IPCC AR4 in Section TS.6 of the Technical Summary of WG I (Solomon et al 2007) is open and frank enough to list out as many as 54 'key uncertainties' as it calls them, in our current understanding of physical climate science. These must be carefully taken into account while making a quantitative assessment of future climate change impacts on India.

For example, according to the IPCC, the key uncertainties about sea level rise are the following: Limitations in ocean sampling imply that decadal variability in global heat content, salinity and sea level changes can only be evaluated with moderate confidence. Global average sea level rise from 1961 to 2003 appears to be larger than can be explained by thermal expansion and land ice melting. Further, existing models cannot address key processes that could contribute to large rapid dynamical changes in the Antarctic and Greenland Ice Sheets which in turn could increase the discharge of ice into the ocean. The sensitivity of ice sheet surface mass balance to global climate change is not well constrained by observations and has a large spread in models. There is consequently a large uncertainty in the magnitude of global warming that, if sustained, would lead to the elimination of the Greenland Ice Sheet.

The uncertainties pertaining to cryosphere are as following: There is no global compilation of in situ snow data prior to 1960. Well-calibrated snow water equivalent data are not available for the satellite era. There is insufficient data to draw any conclusions about trends in the thickness of Antarctic sea ice. Uncertainties in estimates of glacier mass loss arise from limited global inventory data, incomplete area-volume relationships and imbalance in geographic coverage. Mass balance estimates for ice shelves and ice sheets, especially for Antarctica, are limited by calibration and validation of changes detected by satellite altimetry and gravity measurements. Limited knowledge of basal processes and of ice shelf dynamics leads to large uncertainties in the understanding of ice flow processes and ice sheet stability.

About regional projections, these are the key uncertainties: Atmosphere-Ocean General Circulation Models show no consistency in simulated regional precipitation change in some key regions. In many regions where fine spatial scales in climate are generated by topography, there is insufficient information on how climate change will be expressed at these scales.

4.4 Glaciers

By definition, a glacier is a large mass of ice that gets formed by compaction and recrystallization of snow carried downslope under its own weight. This movement, whether rapid or slow, is characteristic of a glacier. Glaciers may form in many different ways, but most commonly because of the wintertime snowfall that has not melted during the next summer. A glacier may have two different zones, one for addition or accumulation of snow, and another for melting or ablation of ice. The lowest extremity of the ablation zone is called the snout of the glacier.

The difference between accumulation and ablation across a hydrological cycle is called the mass balance of the glacier. Whether a glacier is retreating or advancing is determined by this mass balance. The shape and nature of the glacier snout can give an indication of whether a glacier is advancing or retreating. The mass balance of a glacier is influenced greatly by the local climate. In the middle and high latitudes, the hydrological cycle is governed by the annual cycle of air temperature, so that accumulation dominates in winter and ablation in summer. In the tropics, accumulation occurs during the precipitation season, but ablation takes place throughout the year because of warmer temperatures. However in the case of the lofty Himalayan mountains, fresh snow can fall not only in winter but in summer as well, allowing both accumulation and ablation to continue even into the summer season.

4.4.1 Global Warming and the Cryosphere

If there is a change in the climate, a glacier will respond by adjusting its size so as to attain a zero total mass balance, but its response time will depend on where it is located. In the tropical mountains, where ablation goes on all through the year, and the glaciers are steep and shallow, changes in glacier extent may become evident within just a few years of a change in climate. However, the largest glaciers and ice caps of the world, with small slopes and cold ice, may take as long as several centuries to respond to climate change. Therefore, the retreat that is being observed now in some of the larger glaciers may be a response to climate change that had occurred centuries ago and not necessarily attributable to the current phase of global warming.

As per the IPCC AR4 (Lemke et al 2007), strongest specific mass balances or mass losses per unit area have been observed in the glaciers and ice caps of Alaska, northwest U. S., Patagonia and southwest Canada. Because of large areas, the biggest contributions to sea level rise came from Alaska, the Arctic and the Asian high mountains.

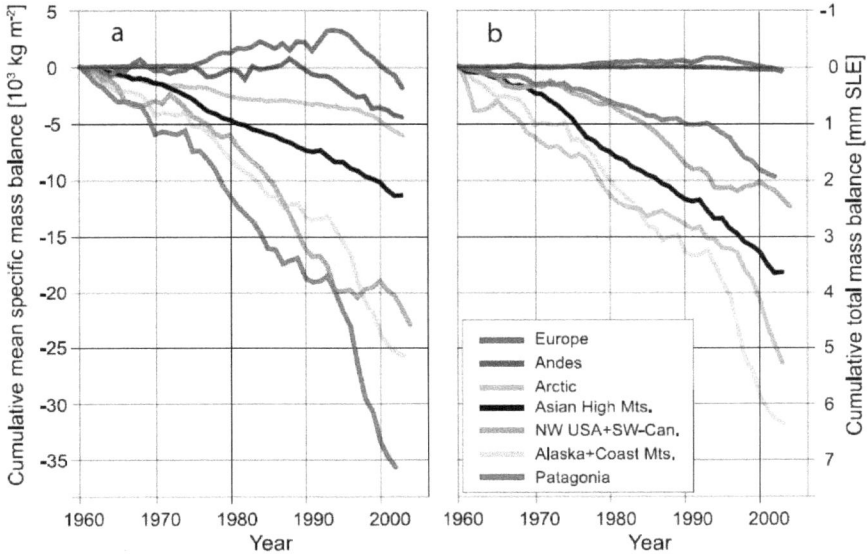

Figure 4.4.1.1 (a) Cumulative mean specific mass balances and
(b) cumulative total mass balances of glaciers and ice caps over large regions
(Source: IPCC AR4, Lemke et al 2007)

Figure 4.4.1.1 shows the change in the cumulative mean specific mass balance from 1960 onwards over different regions which is an indicator of climate change. It also shows the change since 1960 in the cumulative total mass balance which is a measure of the contribution from each region to sea level rise. Most mountain glaciers and ice caps have been shrinking, with the retreat probably having started about 1850, although many northern hemisphere glaciers had a few years of near-balance around 1970, followed by increased shrinkage.

Glaciers and ice caps constitute an extremely small fraction of the earth's cryosphere in terms of both area and volume (Table 4.5.3.1). Their sea level equivalent is just of the order of 0.15-0.37 m as against, for example, 56.6 m of the Antarctic ice sheet. However, what is significant is that during the period 1993-2003, the melting of glaciers and ice caps have contributed to a sea level rise of 0.77 ± 0.22 mm/yr while the Antarctic ice sheet contributed only 0.21 ± 0.35 mm/yr (Table 4.5.3.2). This is why the retreat of glaciers is being regarded as a direct and most visible evidence of global warming.

4.4.2 Himalayan Glaciers

The total area of the globe covered by glaciers is presently estimated to be around 15 million sq km of which 14 million sq km lies in Antarctica and Greenland. Most of the remaining glacier cover is distributed across the northern hemispheric mountain ranges. The Himalayas, including the Karakoram range, constitute the largest glacier system in the world outside of Antarctica and Greenland. There are nearly 10,000 glaciers in the Himalayas, of which 8,000 are in the Indus river basin alone and cover an area of 36,000 sq km. Another 1,600 glaciers are in the Ganga river basin and they cover an area of 4,000 sq km. Within the Indus and Ganga basins also, the glaciers are not evenly spread but about 3,000 glaciers are found in the Chenab basin and 1,200 in the Alaknanda basin (Raina 2009). The largest glacier in the Himalayas is the Siachen glacier which is 73 km long. The second largest is the Gangotri glacier, which is 30 km long and occupies an area of 86 sq km (Jain 2008).

Scientific observations of Himalayan glaciers started about a hundred years ago mainly under the aegis of the Geological Survey of India (GSI). The snout of the Gangotri glacier in Uttarakhand was mapped in detail way back in 1935 and it had shown signs of retreat even at that time. Following the International Geophysical Year of 1957-58 there was an upsurge of activity in the field of glacier monitoring and the GSI extended its snout monitoring to many more glaciers including Siachen, Kumdan, Barashigri, Milam, Pindari, Kalganga, Burphu and Zemu.

It is extremely difficult to employ the traditional method of measuring glacier mass balance over the entire Himalayan range due to inaccessibility and complicated logistics. Until the 1970s, aerial photography was the only remote sensing tool available for this purpose. Thereafter, however, many remote sensing satellites were launched, including India's own Indian Remote Sensing Satellite (IRS) series, and it is became possible to carry out glacier mass balance investigations on a large spatial scale.

Kulkarni et al (2004, 2005 and 2007) have reported the results of several recent investigations made for Himalayan glaciers located in Himachal Pradesh using LISS and WiFS data from successive IRS satellites. The LISS-

IV sensor has three spectral channels, 0.52-0.59, 0.62-0.68 and 0.77-0.86 µ, with a ground resolution of 5.8 m and is therefore useful for mapping small glaciers and ice fields. Normally in the Himalayas, retreat is measured at well-developed and easily accessible valley glaciers. The use of satellite data has made it possible to cover small mountain glaciers as well.

Kulkarni et al (2007) have found that for 466 glaciers in Chenab, Parbati and Baspa basins the mean area of glacial extent has reduced from 1.4 to 0.32 sq km from 1962 to 2001 and that there was an overall reduction of 21 % in the glacial area over this period. However, the loss in glaciated area for large glaciers was 12 % as compared to 38 % for small glaciers. Due to fragmentation, the number of glaciers with higher areal extent has reduced while the number of glaciers with lower areal extent has increased. This indicates that a combination of factors like glacial fragmentation and higher retreat of small glaciers is influencing the sustainability of Himalayan glaciers.

In situ measurements of glaciers are generally difficult because of inaccessible terrain and it is all the more difficult to replicate them in a large number of locations. However, space-based techniques which offer the advantage of a large spatial coverage, are also not without problems. Racoviteanu et al (2008) have reviewed the recent advances in the use of optical remote sensing for mass balance of mountain glaciers, with an emphasis on current algorithms and their limitations in the Himalayas. They have cautioned that the greatest uncertainty in mapping glaciers by remote sensing arises from the presence of debris cover over glacier areas that vitiates the satellite retrievals. Other problems are the limited availability of field validation data such as GPS and specific mass balance measurements and the lack of accurate elevation data for remote glacierized areas.

The retreat of glaciers has assumed great importance in recent years as it is being projected as an indicator of the current global warming. In the case of Himalayan glaciers the issue is not just of academic interest but also of practical concern as India's three major river systems, Ganga, Yamuna and Brahmaputra have their origins in the Himalayas. Because of the farsightedness of the GSI in starting glaciological monitoring more than a century ago, India today possesses a long series of valuable data on Himalayan glaciers which can throw light on the matter of glacier retreat.

A major conclusion that can drawn from this observational evidence is that the Himalayan glaciers have been exhibiting a continuous secular retreat since the earliest recording began in the mid-nineteenth century, and the retreat in the recent years in not unusual. Another interesting and important point is that not all of the Himalayan glaciers have been retreating. For example, the Kangriz glacier has remained unchanged over the last 100 years while the Siachen glacier has not shown any retreat since 1958. The retreat of the Gangotri glacier which had a rate of 20 m/yr until 2000 (Figure 4.4.2.1) has since slowed down considerably, and it is now zero (Raina 2009).

Figure 4.4.2.1 False colour satellite image of the Gangotri Glacier in the Garhwal Himalayas showing its retreat from 1780 to 2001. The blue contours are approximate.
(Source: NASA web site http://earthobservatory.nasa.gov/IOTD)

Making high quality and reliable measurements of glacier retreat at remote and accessible locations in the Himalayas is difficult but is of the greatest need. Otherwise one can very easily arrive at alarmist conclusions. There was a study by Anthwal et al (2006) in which they had presented recession rate data for 19 Himalayan glaciers, presumably collected from various sources. For different Himalayan glaciers, the lengths of their data periods varied between 5 and 121 years, and the actual calendar years were also not the same for all. The recession rates were also quite diverse, ranging from 3 to 36 m/yr. Anthwal et al had mentioned in their study that glaciers in the Himalayas are receding faster than in any other part of the world and, if the present rate continues, the likelihood of them disappearing by the year 2035 is very high.

It is therefore necessary to take a balanced view with regard to the retreat of Himalayan glaciers as has been done by Jain (2008). He has tried to allay fears that the Gangotri glacier has been retreating rapidly as a result of global warming and so the Ganga river will dry up in another 30 years or so, hurting not only the life of millions of Indians but also their sentiments. The Gangotri system is a cluster of glaciers of which the main Gangotri glacier is 30.2 km long, 0.20-2.35 km wide and has an area of 86.32 sq km. Over the past few decades, the rate of recession of the Gangotri glacier has been around 22-27 m/yr (Thayyen 2008) or even as low as 17 m/yr (Kireet Kumar et al 2008). Considering the present length of the glacier, and even assuming a higher recession rate of 40 m/yr, Jain argues that it will take about 700 years to completely melt. Further, there is a likelihood of heavy snowfall in the intervening years which can extend the life of the glaciers.

Another important point made by Jain (2008) is that the Ganga is not totally dependent on glaciers for its water, even in the headwaters region. Most of its catchment area in India is rainfed. Only about 7% of the basin up to Devprayag is glacier-fed. Snow and glacier melt contribute only 29 % to the annual flow at Devprayag; the rest is from rain water. The percentage of snow and glacier-fed area progressively reduces downstream, and so does the contribution. More than 70% of the flow at Haridwar is due to rainfall and four mighty tributaries join the Ganga in Bihar: Ghaghara, Gandak, Kosi and Sone.

Going by the considered views of Jain (2008) and Raina (2009), it is premature at present to come to a definite conclusion that the retreat of the Himalayas glaciers is due to the current global warming. Glaciers are known to be influenced by several geophysical features and local climate fluctuations and it is particularly difficult to correlate individual snout movements to large scale global warming.

The current database about the properties and behaviour of Himalayan glaciers needs to be strengthened and long-term studies on glacier dynamics and mass balance are needed to understand the behaviour of Himalayan glaciers and the impact of climate change on them. It is indeed very encouraging to note that several leading research institutes in India outside the GSI are now engaged in observational and research investigations on the problem of the retreat of Himalayan glaciers. These are Wadia Institute of Himalayan Geology, Dehra Dun; Space Application Centre, Ahmedabad; Snow and Avalanche Study Establishment, Chandigarh; Birbal Sahni Institute of Palaeobotany, Lucknow; National Institute of Hydrology, Roorkee; The Energy and Resources Institute (TERI), New Delhi and several others. It is hoped that these efforts will lead to the emergence of a clearer view of the retreat of Himalayan glaciers vis-à-vis global warming.

4.5 Sea Level

The surface of the earth's ocean is constantly changing. At times it may be turbulent and throw up waves several metres high, while at other times an absolute calm may prevail. There are high tides and low tides. By definition, the average level of the sea surface at a given time relative to a given benchmark or datum is called the sea level. However, measurement of the sea level with a high degree of accuracy is not as easy as it may seem because the sea surface is not uniform and has irregularities and slopes.

Like other substances, water is subject to the process of thermal expansion and this applies as well to the huge water body that constitutes the earth's ocean. Warming of ocean water leads to an expamsion of its volume and therefore a decrease in its density. Cooling of ocean water produces the opposite effect of a shrinking of the volume and an increase in its density. A direct consequenc of the increase or decrease in the total volume of the global ocean water, is a rise or fall of the sea level, and this is termed as eustatic sea level rise or fall. When sea level rise is talked about in the context of global warming, it usually refers to eustatic rise caused by thermal expansion of ocean water.

Isostasy is a geological term that refers to the response of the earth's lithosphere and mantle to changes in the surface loading, such as alterations to the ice or ocean mass, sedimentation and erosion. Vertical isostatic land adjustments take place in order to balance the effect of the load changes. Such local changes may make the sea level appear to have risen or fallen, but this is only a relative sea level rise ot fall with reference to local land movement. When the local impact of sea level rise is to be assessed, it is the isostatic or relative sea level change that assumes importance and this type of sea level change should not be misinterpreted as or confused with sea level rise due to global warming. .

If the values of relative sea level at a place are averaged over an extended period of time, say a month or a year or even longer, the average that is obtained is called the mean sea level. The period of averaging has to be so chosen that the variations due to tides, waves and meteorological factors like wind and pressure get evened out in the averaging process. The elevation of land at a given place is commonly expressed in terms of its height above the mean sea level, and it is a common practice to prominently display it at places like railway stations, airports or tourist spots.

The traditional means of measurement of relative sea level has been the tide gauge. It is only recently that it has become possible to make measurements of the sea surface height using satellite-borne altimeters.

4.5.1 Tide Gauge Measurements

For making an assessment of past global sea level change, tide gauge measurements are the only source of data. While some tide gauge stations, like Mumbai, possess an unbroken data record spanning longer than a century, the conventional tide gauge network as a whole has not been very dense nor were the stations evenly distributed. It was only after the Global Sea Level Observing System (GLOSS) was established in 1990, that the tide gauge network density improved and standardized instruments were installed in the network. The data quality of recent sea level measurements is therefore superior to those made in the first half of the twentieth century and the errors associated with in situ sea level data are falling rapidly. The current GLOSS configuration consists of about 300 tide gauges around the world, but the stations located in the northern hemisphere outnumber those in the southern hemisphere.

Tide gauges of different types and design are deployed at different places as per their application. The conventional tide gauge is a simple instrument that operates basically like a float in a stilling well, and measures the sea level relative to an adjacent geodetic benchmark or datum. However, there is a possibility that this benchmark could itself be affected by vertical land motion of tectonic or other origin, and this introduces a major uncertainty that vitiates tide gauge measurements. The effects of concurrent meteorological factors like barometric pressure and wind speed, have also to be eliminated while deriving the long term averages of sea level change.

There have been several investigations of global sea level rise using historical tide gauge records, but the results have shown a considerable amount of divergence, depending upon the length of the data period and the choice of stations. The estimates of twentieth century global sea level rise have varied between 1.5 and 2 mm/yr (Douglas 2001, Peltier 2001, Miller et al 2004, Holgate et al 2004, Church et al 2004, Church et al 2006). Considering the above results, the Fourth Assessment Report of the IPCC (Bindoff et al 2007) has adopted a rate of 1.7 ± 0.5 mm/yr for the sea level rise in the twentieth century and a slightly higher rate of 1.8 ± 0.5 mm/yr for the sea level rise for the period 1961 to 2003 (Table 4.2.3.1). The interannual variation of global sea level since the year 1870 is shown in Figure 4.5.1.1.

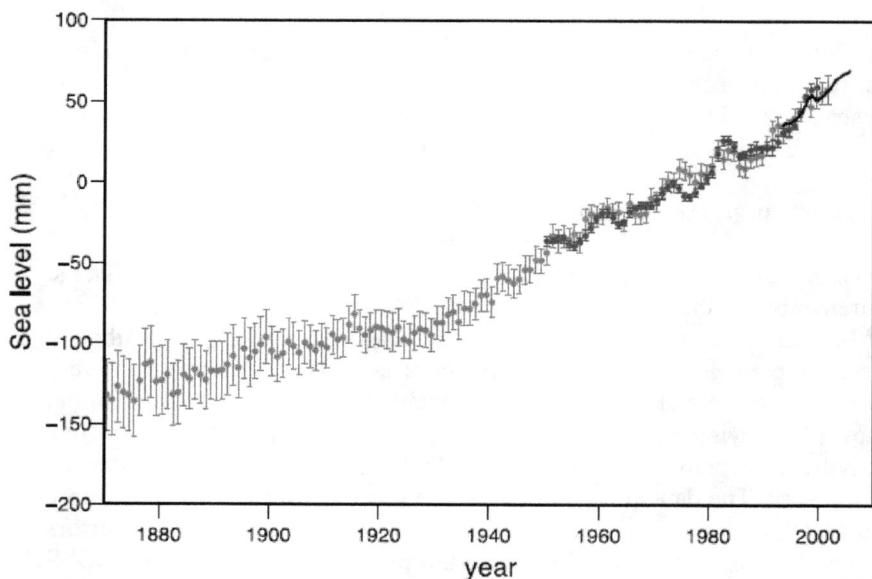

Figure 4.5.1.1 Annual average of the global mean sea level (mm) based upon reconstructed data (red), tide gauge measurements (blue) and satellite altimetry (black). Error bars show 90% confidence intervals.
(Source: IPCC AR4, Bindoff et al 2007)

4.5.2 Satellite Altimetry

Data obtained from satellite-borne altimeters have made a revolutionary impact on the estimation of sea level change. A satellite-borne altimeter is basically a microwave radar operating at a frequency in which the sea surface acts as a good reflector. The radar pulse illuminates a spot on the sea surface which is large enough to smooth the effects of local waves and ripples, and a high pulse repetition rate is used to improve the signal to noise ratio, especially when the ocean surface is rough. The two-way travel time gives the height of the satellite above the sea surface. It must be remembered here that the aim is to detect changes in the sea level of the order of millimetres from a satellite altitude of several hundred kilometres. The process demands extreme precision and the satellite position itself has to be tracked by a global network of ranging stations. Corrections have to be applied to the travel time of the pulse to account for ionospheric and atmospheric delays, and known tidal corrections are made as well. The sea surface height is calculated as the difference between the distance of the satellite from the centre of the earth and the distance between the sea surface and the satellite as measured by the altimeter.

The slope of the altimeter return pulse is stretched in time because of the delay between reflections from the wave crests and troughs. This helps to estimate the significant wave height which itself is useful in correcting the sea surface height observations. The strength of the return pulse is an indicator of the wind speed. The altimeter is the only instrument that can provide real time information on the sea state and the waves generated by cyclonic storms.

A vital advantage of satellite-borne altimeters over tide gauges is that the satellite-determined sea surface height is not affected by vertical land movements as in the case of tide gauge observations. Moreover, satellite estimations are so plentiful compared to the conventional data that it becomes possible to make an analysis of not only global but regional sea level rise as well.

However, altimeter measurements are not without their own errors, some of which may arise from the instrument itself. Other errors could be attributed to the fact that the actual speed of propagation of electromagnetic radiation in the atmosphere may not be the same as that used in the computations. Additional errors may creep in due to ocean tides, inverse barometer effects, and inaccuracies in the modelling of the geoid itself (Ali 2003).

The first successful radar altimeter mission was Geosat, which was launched by the U.S. Navy in 1985 and worked for more than four years. The real breakthrough however came with TOPEX/Poseidon, launched in August 1992 as a joint NASA-CNES programme. The satellite was positioned at an altitude of 1336 km, in a circular, non-sun-synchronous, 66° inclination orbit covering 95 % of the world's ice-free oceans between 66°N and 66°S latitudes with a 10-day repeat cycle. The TOPEX/Poseidon mission lasted about 14 years, until the satellite was de-commissioned on 18 January 2006. TOPEX/Poseidon resulted in the generation of the longest ever and most complete direct observations of sea level change, including the El Nino event of 1997.

The range measurements after correction for atmospheric and instrumental effects, were accurate to 3-4 cm. The range measurements when subtracted from estimates of the satellite orbital height, resulted in ocean height measurements of 4-5 cm accuracy relative to the centre of the earth. This accuracy figure pertained to a spot on the ocean surface of the size of a few kilometres directly beneath the TOPEX/Poseidon satellite. By averaging the thousands of measurements collected by the satellite in the 10 days time that it took to cover the global oceans, global mean sea level could be determined with a precision of the order of millimetres.

In December 2001, the Jason-1 satellite, another joint NASA-CNES venture, was launched as a replacement of the TOPEX/Poseidon mission and placed in the same orbit with similar instruments. It is continuing to make the same measurements. Parallelly, on 1 March 2002, ESA's ENVISAT satellite was launched with another radar altimeter in an orbit repeating at 35 days instead of 10 days for Jason and sampled the ocean on a smaller spatial scale, complementing the Jason data.

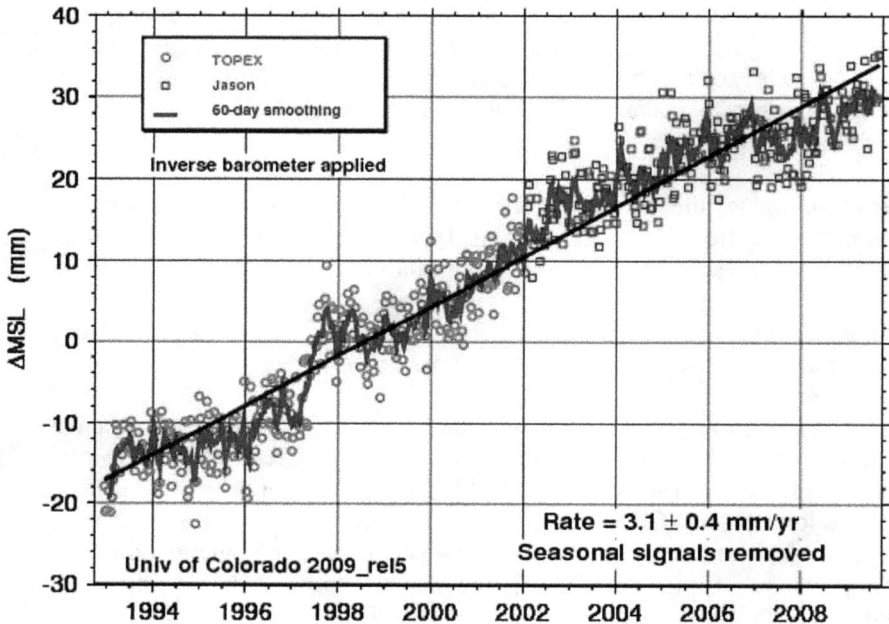

Figure 4.5.2.1 Rate of global sea level rise as monitored by satellite-borne altimeters from 1992 to 2009 (Source: University of Colorado at Boulder web site http://sealevel.colorado.edu)

Since 1992, high precision global sea level measurements are being made continuously by successive satellite missions, first by TOPEX/Poseidon, from 2001 by Jason-1, and presently by Jason-2. Over this period, the satellite-estimated average global sea level has shown a rising trend of 3.1 ± 0.4 mm/yr (Figure 4.5.2.1) which is considerably higher than the trend obtained from tide gauge data (Leuliette et al 2004). It is unknown whether the higher rate in 1993 to 2003 is due to decadal variability or an increase in the longer-term trend.

Using space gravimetry observations from Gravity Recovery and Climate Experiment (GRACE), Cazanave et al (2009) showed that sea level rise in recent years can be mostly explained by an increase of the mass of the oceans. The increase in mass increase since 2003 is due to equal contributions from melting of polar ice sheets and mountain glaciers. The total ocean mass contribution is about 2 mm/yr over the years 2003-2008 which represents about 80% of the altimetry-based rate of sea level rise over that period. The result agrees well with the Argo-based value estimated as 0.37 mm/yr over 2004-2008.

The joint U. S. - Europe Ocean Surface Topography Mission (OSTM/Jason-2) has been launched in June 2008. This will extend the time series of ocean surface topography measurements beyond TOPEX/Poseidon and Jason time frames to accomplish two decades of observations. Another altimeter called AltiKa, working in Ka-band at a frequency of 35 GHz, is likely to be launched shortly aboard the Indian Oceansat-3 satellite as a joint India-France mission and will complement Jason-2.

4.5.3 Global Warming and Sea Level Rise

A direct consequence of global warming is a rise in the global sea level. This will happen for three different reasons. First, the heating of the ocean will cause its thermal expansion and the volume of ocean water will increase, producing an eustatic sea level rise. Second, melting of the ice sheets over the polar regions, particularly the massive ones such as those over Greenland and western Antarctica, will add water to the adjacent ocean. Third, melting of mountain glaciers and small ice caps will result in the conversion of terrestrial water storage from solid to liquid phase, and create increased surface runoff into rivers and eventually into the ocean.

The earth's cryosphere stores about 75 % of the world's freshwater. This includes snow on land, sea ice, glaciers and ice caps, ice shelves, the Greenland and Atlantic ice sheets, seasonally frozen ground and permafrost. About 10% of the earth's land surface is permanently covered by ice, and this is mostly in the form of ice sheets over Greenland and Antarctica. The ice that lies in ice caps and glaciers outside Antarctica and Greenland is negligible in comparison. On a regional scale, however, many glaciers and ice caps are important for freshwater availability. About 7% of the ocean is covered ny sea ice. In winter, about half of the northern hemispheric total land area gets covered by snow.

Assuming that the earth's cryospheric components were to completely break down and melt into liquid water on account of global warming, the sea level rise that would occur as a consequence is called their sea level equivalent. This gives an indication of their relative contributions to sea level rise as a resilt of global warming. The sea level equivalent is computed by converting their area in million km^2 and volume in million km3 to a sea level rise in metres. Table 4.5.3.1 shows the area, volume and sea level equivalents of the different cryospheric components of the earth. The Greenland and Antarctic ice sheets have by far the highest potential for sea level rise which is of the order of 7 m and 57 m respectively. For all other components, the potential sea level rise is close to zero or of the order of some cm. These figures are based on the assumption that an ice sheet mass loss of 100 Gt/yr is equivalent to 0.28 mm/yr of sea level rise. A global isostatic anomaly (GIA) correction was applied to observations from both tide gauges and satellite altimetry (IPCC AR4, Bindoff et al 2007).

The sea level equivalent is an extreme or ultimate value. The rate of sea level rise that has been actually observed in recent times is, however, several orders of magnitude smaller (Table 4.5.3.2) than the sea level equivalent. For example, the Antarctic ice sheet has been contributing just about 0.1 to 0.2 mm/yr as against its sea level equivalent of 57 m, i.e., 57000 mm.

Another interesting feature of Table 4.5.3.2 is that for the 1961-2003 period, the sum of all known contributions from sources related to global warming is smaller than the observed sea level rise. The difference which remains unexplained could perhaps be attributed to past climate change effects. For the period 1993-2003, however, for which the observing system is much better, the contributions from thermal expansion (1.6 ± 0.5 mm/yr) and loss of mass from glaciers, ice caps and the Greenland and Antarctic ice sheets together make up for a sea level rise of 2.8 ± 0.7 mm/yr. The unaccounted difference is much smaller, suggesting that the global warming contributions to sea level rise are now much more dominant than before.

As per the IPCC AR4, global sea level is projected to rise during the 21st century at a greater rate than during 1961-2003. By the end of the century, under the A1B emissions scenario, global sea level may be 22 to 44 cm higher than the 1990 level, and continuing to rise at a rate of about 4 mm/yr (Figure 4.5.3.1). Regional sea levels may be different from the global mean by -15 to +15 cm.

Table 4.5.3.1 Area, volume and sea level equivalent of the earth's cryospheric components
(Source: IPCC AR4, Bindoff et al 2007 and Lemke et al 2007)

Cryosphere component	Area (million km^2)	Ice volume (million km^3)	Sea level equivalent or potential sea level rise (m)
Snow on land*	1.9-45.2	0.0005-0.005	0.001-0.01
Sea ice*	19-27	0.019-0.025	~0
Glaciers and ice caps**	0.51-0.54	0.05-0.13	0.15-0.37
Ice shelves	1.5	0.7	~0
Greenland ice sheet	1.7	2.9	7.3
Antarctic ice sheet	12.3	24.7	56.6
Seasonally frozen ground*	5.9-48.1	0.006-0.065	~0
Permafrost	22.8	0.011-0.037	0.03-0,1
*Values indicate annual range of variation **Values indicate range of estimates			

Table 4.5.3.2 Observed global sea level rise during 1961-2003 and 1993-2003 compared with components of sea level rise attributable to sources related to global warming
(Source: IPCC AR4, Bindoff et al 2007 and Lemke et al 2007)

Source	Global sea level rise (mm/yr)	
	1961-2003	1993-2003
Thermal expansion of ocean	0.42 ± 0.12	1.6 ± 0.5
Glaciers and ice caps	0.50 ± 0.18	0.77 ± 0.22
Greenland ice sheet	0.05 ± 0.12	0.21 ± 0.07
Antarctic ice Sheet	0.14 ± 0.41	0.21 ± 0.35
Sum	1.1 ± 0.5	2.8 ± 0.7
Observed	1.8 ± 0.5	3.1 ± 0.7
Difference (Observed – Sum)	0.7 ± 0.7	0.3 ± 1.0

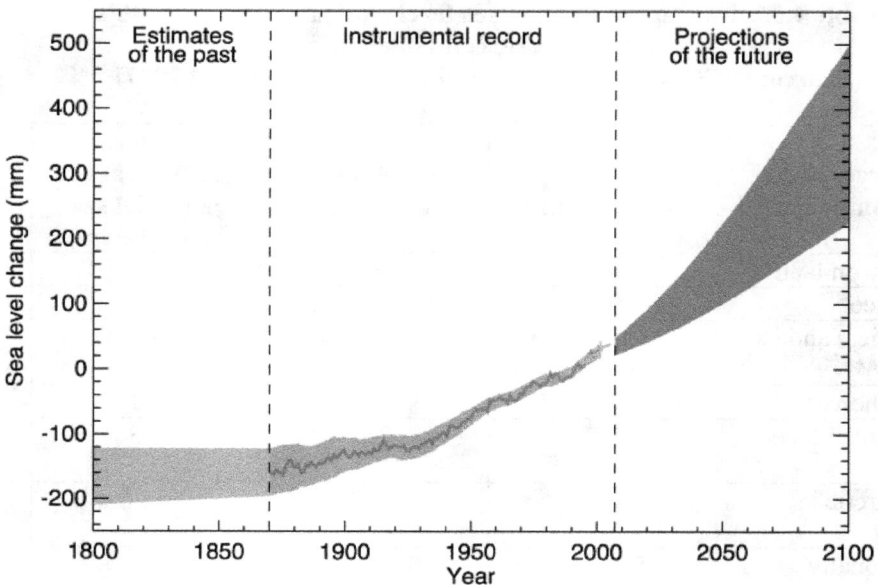

Figure 4.5.3.1 Change in global average sea level rise relative to the present. Gray shaded area is an estimation for 1800-1870. Red line shows measurements by tide gauges and the short green line shows satellite measurements. Red shaded area indicates the uncertainty in tide gauge data. Blue shaded area represents the range of model projections for a medium growth emissions scenario IPCC SRES A1B.
(Source: IPCC AR4, Bindoff et al 2007)

Thermal expansion is projected to contribute to more than half of the sea level rise. However, the projections have an important uncertainty about them which arise from the possibility that the actual rate of discharge from the ice sheets could be quite different from what is assumed in the models. While land ice will continue to lose mass increasingly rapidly as the century progresses, IPCC AR4 cautions that quantitative projections of the consequences of accelerated ice flow as observed in recent years, cannot be made with confidence, owing to limited understanding of the relevant processes.

4.2.4 Sea Level Rise along the Indian Coastline

What is not often realized is that sea level rise along different segments of the vast Indian coastline has not been and is not expected to be uniform. The sea level at a given place is influenced by several local factors such as coastal ocean temperature, salinity, wind, atmospheric pressure and ocean currents. The sea level is also affected by the changes in coastal geometry resulting from sedimentation, coastal erosion, storm surges and the action of waves (Das 2001).

Levelling observations conducted during the Great Trigonometrical Survey of India (1858-1909) and subsequent observations have shown that mean sea level along the east coast of India is higher than that along the west coast, the difference in sea level between Visakhapatnam and Mumbai being about 30 cm. Shankar et al (2001), using simulations with a 1.5-layer reduced gravity model have attributed 60% of this sea level difference to the mean, large-scale wind-forced circulation and 40% to the alongshore gradient in salinity.

As already mentioned, sea level trends derived from historical tide gauge records are sensitive to the choice of stations and the data period analyzed. Das et al (1991) analyzed the tide gauge data for four Indian stations selected for their reliability and found that there was no evidence of a monotonic rising trend at any of them. The Mann-Kendall test revealed a rising trend at Mumbai from 1940 to 1986 and at Chennai from 1910 to 1933 and none of the other records showed a significant trend.

Unnikrishnan et al (2006) using tide gauge data up to 1992 found that the sea level showed an increasing trend for Mumbai, Visakhapatnam and Kochi but that it showed a decreasing trend for Chennai. In a subsequent study, Unnikrishnan et al (2007) made an extensive analysis of the mean sea level data from coastal tide gauges in the north Indian Ocean and showed that the observed low frequency variability of sea level was consistent among the stations within the basin. Statistically significant trends obtained from records longer than 40 years yielded sea level rise estimates between 1.06 and 1.75 mm/yr, with a regional average of 1.29 mm/yr, when corrected for GIA using model data.

Figure 4.5.4.1 shows the interannual variation of monthly mean sea level derived from tide gauge measurements and the 5-month running average for three coastal stations on the east and west coasts of India. In these plots, the average seasonal cycle and linear sea level trend have been removed. In these figures, solid vertical lines if present, indicate times of any major earthquakes in the vicinity of the station and dashed vertical lines mark questionable data. The results are also summarized in Table 4.5.4.1.

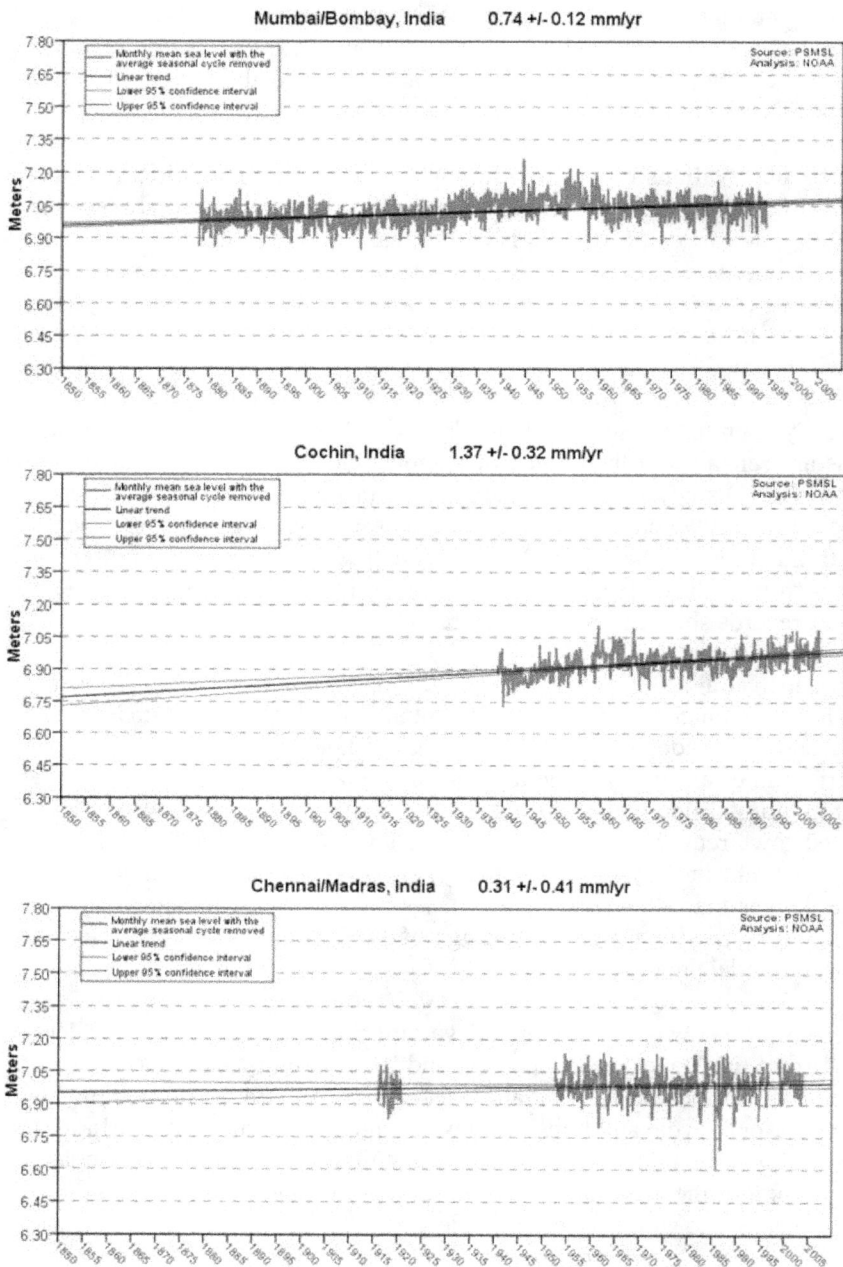

Figure 4.5.4.1 Long term monthly mean sea level data series for Mumbai, Kochi and Chennai (Source: U. S. Environmental Protection Agency web site http://tidesandcurrents.noaa.gov/sltrends/sltrends.html)

**Table 4.5.4.1 Sea level rise for selected stations along the Indian coast
(Source: Same as for Figure 4.5.4.1)**

Station	Period of analysis (years)	Trend in sea level rise (mm/yr)	95 % confidence interval (mm/yr)	Sea level rise per century (cm)
Mumbai	1878–1994	0.74	± 0.12	7.4
Kochi	1939–2004	1.37	± 0.32	13.7
Chennai	1916-2003	0.31	± 0.41	3.1

4.5.5 Sea Level Rise over the Indian Ocean

The prime advantage of satellite altimetry over tide gauge measurements is that satellite-derived sea level data are available across the oceans and not just on the coast. Although the TOPEX/Poseidon sea level data time series from 1992 onwards shows (Figure 4.5.4.1) a linear trend of 3.1 mm/yr in the global average sea level, the spatial distribution of sea level trends in the Pacific and Indian Ocean is significantly different from the global trend. It is very interesting to see, and of great importance to India, that over the Arabian Sea and the Bay of Bengal, during the recent period 1992-2009, the sea level rise has been negligibly small and over many parts the sea level has in fact exhibited a fall of the order of 3 mm/yr. In comparison, over eastern Indian Ocean and western Pacific Ocean, the sea level rise has been very significant, as high as 15 mm/yr in some parts. (Figure 4.5.5.1). Similar results have been derived by Church et al (2006), who have attributed them to large scale interannual variability in general and to the La Nina conditions that prevailed in 2001 in particular. Partially, the regional variations in sea level trends can also be attributed to changes in the thermal structure of the oceans, which in turn are linked to changes in surface heating effects and ocean circulation (Bindoff et al 2007).

Thompson et al (2008) investigated the long term variability of temperature and sea level over the north Indian Ocean during the period 1958-2000 using an Ocean General Circulation Model. The model simulated fields were compared with the sea level observations from tide gauges, TOPEX/Poseidon satellite, and in situ temperature profile and sea surface temperature observations from moored buoys. They found that long warming episodes in the SST over the north Indian Ocean were followed by short episodes of cooling. The model temperature and sea level anomaly over the north Indian Ocean showed an increasing trend in the study period. The thermosteric component of the north Indian Ocean sea level anomaly showed a linear increasing trend of 0.31 mm/year.

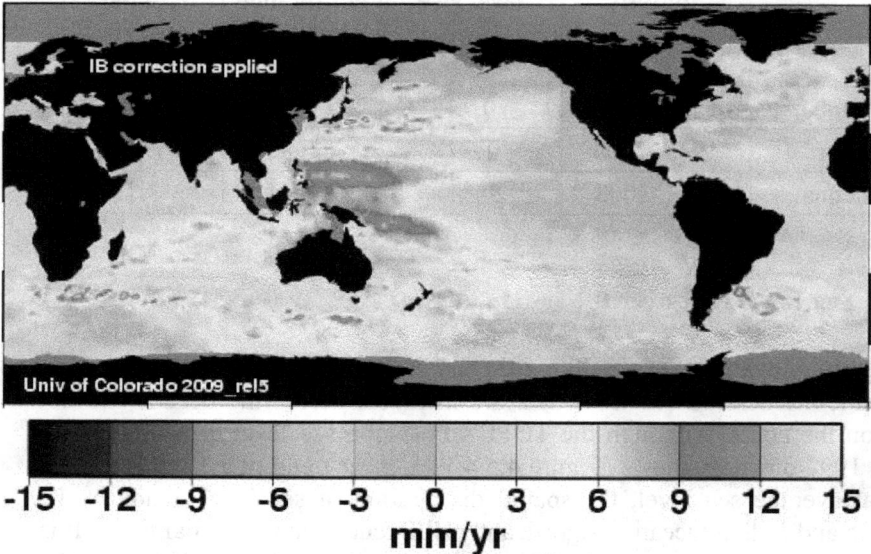

Figure 4.5.5.1 Trends in sea level change over the global oceans derived from TOPEX/Poseidon altimeter data for 1992-2009. Positive (negative) values denote rise (fall) in sea level.
(Source: University of Colorado at Boulder web site http://sealevel.colorado.edu)

4.5.6 Impacts of Sea Level Rise

A very comprehensive analysis of the impacts of likely sea level rise on 84 developing countries has been carried out for the World Bank by Dasgupta et al (2007). It starts with the premise that sea level rise due to global warming is estimated to be of the order of 1 to 3 m in this century, while in the event of an unexpectedly rapid breakup of the Greenland and West Antarctic ice sheets the sea level rise could be as much as 5 m. Dasgupta et al have assessed the consequences of inundation that may occur in this range of scenarios using GIS techniques on different impact elements such as land, population, agriculture, urban extent, wetlands, and GDP. The assessment has been made for each of the 84 countries individually as well as for larger geographical regions comprising groups of countries. The results reveal that the sea level rise caused by global warming could lead to the displacement of hundreds of millions of people in the developing world and produce overwhelming agricultural, ecological, economic and other impacts. However, a most significant result that emerges from the study of Dasgupta et al is that the impacts vary drastically from country to country. The countries that are likely to be the worst affected are Vietnam, A. R. of Egypt, and the Bahamas, and for many others, including China, the absolute magnitudes of potential impacts are very large. At the other extreme, many developing countries, including India, may experience very limited impacts.

Table 4.5.6.1 Impact of 1 m sea level arise on various factors in different countries (Source: Dasgupta et al 2007)

	Land area %	Popu-lation %	GDP %	Urban Extent %	Agri-cultural Extent %	Wet-lands %
Vietnam	5	11	10	11	7	29
Bahamas	12	4	5	4	4	18
Egypt	<1	9	6	6	13	8
China	<1	2	2	1	<1	<1
Bangladesh	1	1	<1	<1	<1	1
Pakistan	<1	<1	<1	<1	<1	7
Sri Lanka	<1	<1	<1	1	<1	2
India	<1	<1	<1	<1	<1	1
India 5 m	1	3	3	2	<1	5

With a 1 m sea level rise (Table 4.5.6.1), Vietnam, Bahamas and Egypt rank among those countries which would bear the maximum impact. Bahamas is by far the most impacted country, with close to 12 % of its area affected. Around 11 % of Vietnam's population and 9 % of Egypt's population would be impacted. Vietnam could also see 10 % of its GDP and 11 % of urban extent impacted. Egypt's agricultural extent would experience the largest percentage impact of 13 %. Nearly 29 % of Vietnam's wetlands would be impacted.

Compared to the other regions of the world, sub-Saharan Africa and South Asia would bear a lesser impact. Within the countries of south Asia, India would be the least impacted. This is a very significant and reassuring result brought out by Dasgupta et al (2007). For a 1 m sea level rise, which may take much more than a century to happen, India is likely to have an impact of just 1 % or less on all the six fronts. Even if there is a 5 m sea level rise perhaps due to a catastrophic breakup of the Greenland and Antarctic ice sheets, which is most unlikely to happen, the adverse impacts on India will be very limited (Table 4.5.6.1 bottom row).

From Figure 4.5.5.1 also it has been observed that the sea level of the Arabian Sea, Bay of Bengal and the western Indian Ocean has not been rising but in fact been falling in recent years.

However, there is no room for complacency in this regard on the part of India as the coastal zone is of crucial importance to India on several counts. The metropolises of Mumbai, Chennai and Kolkata and several smaller cities are situated in the coastal belt. Many of India's vital and strategic installations in the fields of atomic energy, space and defence are located right on the coast. There are 11 major ports and 130 minor ports besides numerous smaller natural harbours. Both the eastern and western coasts are regularly visited by tropical cyclones and storm surge is a major factor in the loss of life. India has therefore to consider the threat of sea level rise carefully and seriously and put appropriate plans in place for reducing its impacts.

The Indian mainland has a 5,700 km long coast line, and if the Lakshadweep islands in the Arabian Sea and the Andaman and Nicobar islands in the Bay of Bengal are included, the length of the coastline is as much as 7,500 km, An important consideration here is that the observed sea level rise is not uniform along the entire coastline and not all of it is due to global warming alone. Even the mean sea level along the coast is higher over the Bay of Bengal than over the Arabian Sea, there being a difference of 30 cm between Visakhapatnam and Mumbai as shown by observations. This can possibly be attributed partly to the mean large-scale wind-forced monsoon circulation and partly to the alongshore salinity gradient (Shankar et al 2001).

There is, however, a growing tendency to attribute all sea level rise observed anywhere on the coast entirely to global warming, which needs to be curbed and a balanced view taken considering other possible reasons as well. Problems such as erosion, flooding, subsidence, salinization and deterioration of local ecosystems like mangroves already prevail in our coastal regions and climate change may cause them to worsen (Pavri 2009). Anthropogenic factors also play a role in the deterioration of the environment of the coastal zone, through illegal housing construction, setting up of industries, water pollution and so on. Sea level rise may not by itself appear to be\\pose a major threat, but it could aggravate such pre-existing problems and lead to increased coastal inundation, loss of land and particularly agricultural land, migration of population and beach erosion with an adverse impact on tourism.

Different sectors of the long Indian coastline have different physiographic and environmental characteristics. The western coast adjoining the Arabian Sea has a wide continental shelf with backwaters and mud flats. The eastern coast encompassing the Bay of Bengal has mangroves, deltas and tidal creeks. Islands have coral reefs in their vicinity. The northeast coast of West Bengal, Orissa and north Andhra Pradesh is an emerging coastline with no offshore bar. The southeast coast of south Andhra Pradesh and Tamil Nadu is also an emerging coastline but with an offshore bar. On the east coast, the shorelines off the mouths of Ganga, Mahanadi, Krishna, Godavari and Cauvery rivers are neutral and highly dynamic due to the large influx of sediments. The west coast has a submerging coastline. The above local features make certain sectors of the Indian coastline more vulnerable to sea level rise than others. Some of the highly vulnerable areas are the Sunderbans, the Hooghly estuary, the mouths of rivers on the east and west coasts, Konkan, Goa. Kerala, and the Gulfs of Mannar, Kutch and Khambat.

To tackle the likely impacts of sea level rise resulting from global warming, what is needed therefore is a comprehensive coastal zone management plan to be put in place. The coastal zone should by definition extend on either side of the coastline and include the sea bed and inland water bodies affected by tides. Management of the coastal zone should cover ecologically and culturally significant resources taking into account the vulnerability to natural hazards. If such a plan is adopted and a mechanism for its implementation put into operation, it would go a long way towards the conservation of coastal and marine areas, protection of the coastal populations, opportunities for better livelihood, and sustainable development of the coastal zone.

4.6 Tropical Cyclones

Tropical cyclones over the Bay of Bengal and the Arabian Sea are much fewer in number than those in other ocean basins, but they have been responsible for comparatively much heavier losses of life and property in the countries of the south Asian region. In 1970, a severe cyclonic storm that made landfall in Bangladesh caused 300,000 deaths and was one of the most deadly natural disasters in modern history. Again in 1991, Bangladesh was struck by a severe tropical cyclone in which 138,000 people perished. In India, thousands of people have lost their lives in the states of Orissa and Andhra Pradesh which are most vulnerable to tropical cyclones, the latest example being that of the Orissa supercyclone of 1999 in which the death toll was 10,000. It is therefore natural for India to be concerned about whether and how the pattern of frequency and intensity of tropical cyclones is likely to be altered because of global warming.

One of the pre-conditions for the formation of a tropical cyclone over the ocean is that the sea surface temperature (SST) should be 26.5 °C or higher. Tropical storms therefore tend to form only over certain ocean basins of the world and in certain preferred seasons where and when there is a possibility of this condition being satisfied in the first place. The primary effect of global warming, logically speaking, would be to cause the climatological SST isotherm of 26.5 °C to spread out and so favour the formation of tropical cyclones over a larger oceanic area than at present.

However, it should be remembered here that a warm ocean is just one of the many pre-conditions for the formation of a tropical cyclone, and not the only one. Tropical storms do not have an independent existence of their own but they are embedded in the general atmospheric flow which exerts an influence over them and which they may even modify. Further, it is not only the number of tropical storms that is important, but also the peak intensity that they reach, and the length and orientation of the tracks that they follow. Hence, statistical correlations between global warming and the frequency of occurrence of tropical storms cannot be derived or viewed in isolation without due regard to these other aspects.

4.6.1 Historical Data on Tropical Cyclones

Any statistical investigation of tropical cyclones, whether it is in the context of global warming or not, encounters two basic difficulties, the first one being the small size of the statistical sample. Over the north Indian Ocean, for example, just 4 or 5 tropical cyclones develop in a year on an average and within them those over the Arabian Sea are still rarer than those over the Bay of Bengal. The number of storms reaching the supercyclone stage is extremely small. Even over the Atlantic and Pacific Ocean basins, category 5 hurricanes form only a small percentage of all tropical storms. Table 4.6.1.1 gives details of the categorization of hurricanes.

The second problem is the non-homogeneity of the data base. Well-documented historical records of tropical cyclone tracks and intensity for the Bay of Bengal and Arabian Sea compiled by the India Meteorological Department (IMD) date back to more than a century. Since getting caught in a tropical cyclone was the most dangerous of all hazards faced by shipping vessels, mariners and ship captains maintained meticulous weather logs particularly about tropical storms. When the ships berthed at ports, meteorologists retrieved this vital information and used it for updating their records. Hence, even when there were no upper air observations, no weather radars and no meteorological satellites, reliable records of tropical cyclones could be generated and they are still used for reference and climatological investigations. The situation regarding historical storm data is similar over other occur basins too.

Table 4.6.1.1 Categorization of Hurricanes

System Features	Category				
	1	2	3	4	5
Sustained winds (m/s)	33-42	43-49	50-58	59-69	≥70
Storm surge (m)	1.2-1.5	1.8-2.4	2.7-3.7	4.0-5.5	≥5.5
Central Pressure (hPa)	980	965-979	945-964	920-944	<920

The most significant advance in the monitoring of tropical cyclones came in the 1960s with the advent of polar orbiting weather satellites that could actually observe a cyclone from space. The satellite pictures however required human interpretation which was a highly subjective process. Dvorak (1975) developed a technique of tropical cyclone intensity estimation which was largely objective and it came into use globally on an operational basis with some subsequent modifications (Dvorak 1984, Dvorak et al 1992). By the 1980s, geostationary satellites had entered the scene and tropical cyclones could be monitored every half hour or at more rapid intervals. It was not always possible to use such an enormous volume of data available in real time, and a lot of information including damage reports would come in after the event. It became the practice of operational meteorologists to make a review of all such relevant data after the cyclone season was over and come up with a best track and intensity estimate that would go in the final climatological records. In such post-season reviews, considerations of temporal consistency, accuracy of point values and climatology were often in conflict and a compromise fix had to be made.

Dvorak's technique enabled an estimation of cyclone intensity in terms of both maximum winds and central pressure. However, it assumed that the winds and pressures are consistent, which may not always be the case. Smaller tropical cyclones in particular can have stronger winds for a given central pressure than a larger tropical cyclone with the same central pressure. Another shortcoming of Dvorak's technique was that it was calibrated statistically with measurements made during aircraft reconnaissance flights over the Atlantic and northwest Pacific basins. These statistical correlations do not strictly apply to other ocean basins like the north Indian Ocean but they have to be accepted in the absence of aircraft measurements there. It is only in recent years that supplemental information has become available in real time from satellites like the Tropical Rainfall Measuring Mission (TRMM) which can scan tropical cyclones with an onboard precipitation radar, and QuikSCAT which carries a scatterometer for measuring cyclone winds at the sea surface. Such data are of great help in overcoming the shortcomings of the statistical Dvorak technique.

Landsea et al (2006) have highlighted the problem of historical databases of tropical cyclone intensity and number. To illustrate their point, they have reapplied the Dvorak technique retrospectively to infrared satellite images of five tropical cyclones that developed over the North Indian Ocean basin during the period 1977-1989. According to Landsea et al, these storms which were earlier classified as the equivalent of category 3 or lower, need to be upgraded in intensity in hindcast. If such a revision is indeed accepted, it would wipe out any increasing trend in the frequency of more intense storms noticeable in recent years over the north Indian Ocean basin.

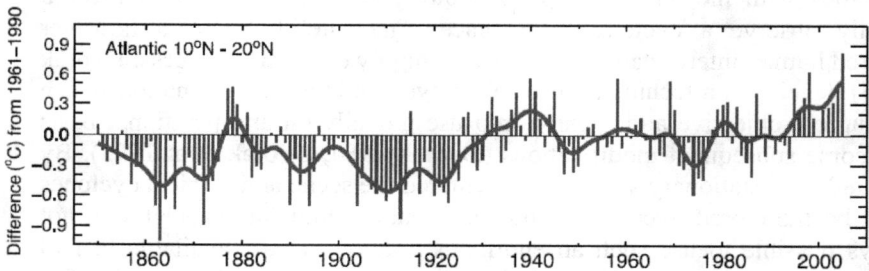

Figure 4.6.2.1 Annual sea surface temperature anomalies (°C) relative to the 1961-1990 mean in the region of hurricane formation over the north Atlantic Ocean. (Source: IPCC AR4, Solomon et al 2007)

4.6.2 Global Warming and North Atlantic Hurricanes

As per IPCC AR4 (Trenberth et al 2007) there is observational evidence for an increase of intense tropical cyclone activity in the North Atlantic Ocean since about 1970, which correlates well with the increase in SST over that region (Figure 4.6.2.1). There are also suggestions of increased intense tropical cyclone activity in some other regions where concerns over data quality are greater. Estimates of the potential destructiveness of tropical cyclones also suggest a substantial upward trend since the mid-1970s, with a trend towards longer lifetimes and greater intensity. However, the frequency and intensity of tropical cyclones are known to exhibit natural variability on the decadal scale, ENSO being one of its causes. It is therefore difficult to isolate very clearly the influence of global warming on tropical cyclone activity.

In 2004, four hurricanes hit Florida in the U. S. and as many as ten typhoons hit Japan. The year 2005 was again marked by a record-breaking number of seven major hurricanes that threatened or directly affected the United States in succession and they were associated with high SST anomalies in their formation region of the north Atlantic Ocean. The estimated economic damage in the U. S. from the 2004 Atlantic hurricane season was nearly $ 40 billion. In 2005, the unprecedented damage caused by Hurricane Katrina alone was estimated to have been more than $ 100 billion and 1300 human lives were lost. As a result, an intense scientific debate began in 2005 about whether the U. S. was getting increasingly vulnerable to stronger hurricanes and rising sea levels associated with the emission of greenhouse gases and global warming. The debate had other aspects too, and the Bulletin of the American Meteorological Society carried an article in its August 2006 issue which was interestingly titled as "Mixing politics and science in testing the hypothesis that greenhouse warming is causing a global increase in hurricane intensity" (Curry, Webster and Holland 2006). This debate has since been going on inconclusively.

The issue whether the increase in north Atlantic hurricane activity since 1995 could be attributed to global warming was in fact raised by Trenberth (2005) before the 2005 hurricane season had begun. His argument was that even if it was difficult to find a statistical trend in a parameter that had small values and high variability, it did not prove that there was no trend. All that it meant was that the particular data sample was insufficient to prove it and more data was required.

Trenberth stated that since 1995 the Accumulated Cyclone Energy (ACE) index for all but two Atlantic hurricane seasons had been above normal except in the El Nino years of 1997 and 2002. The hurricane seasons from 1995 to 2004 averaged 13.6 tropical storms, 7.8 hurricanes, and 3.8 major hurricanes, and the ACE index was 169% of the median. In contrast, the hurricane seasons during the previous 25-year period from 1970 to 1994 averaged 8.6 tropical storms, 5 hurricanes, and 1.5 major hurricanes, and the ACE index was 70% of the median. In 2004, ACE reached the third highest value since 1950; there were 15 named storms, including nine hurricanes.

As regards SST, Trenberth found that in addition to interannual and multidecadal variability, there was a nonlinear upward trend over the 20th century. This trend was most pronounced in the past 35 years in the extratropical north Atlantic and it is associated with global warming and can be attributed to human activity. In the tropical north Atlantic, the region of hurricane formation, multidecadal variability dominated SSTs, but the 1995-2004 decadal average was highest on record. Other factors that have influenced the increase in hurricane activity in the past decade include an amplified high-pressure ridge in the upper troposphere across the central and eastern north Atlantic, reduced vertical wind shear and African easterly lower atmospheric winds that favour the development of hurricanes from tropical disturbances moving westward from the African coast.

Trenberth concluded that trends associated with human influences are evident in the environment in which hurricanes form, and our physical understanding suggests that hurricane intensity and associated rainfall are probably increasing, even if this increase cannot yet be proven with a formal statistical test. The fact that the number of hurricanes have increased in the Atlantic is no guarantee that this trend will continue. The key scientific question is not whether there is a trend in hurricane numbers and tracks, but rather how hurricanes are changing.

Trenberth's work was followed by a paper by Emanuel (2005) on what he called the increasing destructiveness of tropical cyclones over the past 30 years. His contention was that it did not matter whether or not there was a detectable trend in the global annual frequency of tropical cyclones, as it was not by itself an optimal measure of a growing tropical cyclone threat. Instead he defined an index of the potential destructiveness of hurricanes and showed that this index had increased markedly since the 1970s. Emanuel's Power Dissipation Intensity (PDI) index is the integral over the life of each tropical storm of its maximum surface wind speed cubed, also accumulated over each year. PDI is thus a combined measure of Atlantic hurricane frequency, intensity, duration and surface area covered. Emanuel found a strong correlation, on multi-year time scales, between local tropical Atlantic SST and PDI, and he also found some evidence that PDI levels in recent years are higher than in the previous active phase of Atlantic hurricanes in the 1950s and 1960s. If the increase in Atlantic SST is attributable to global warming, then the PDI is indirectly attributable to anthropogenic GHG-related warming. Since the north Atlantic SST is projected to rise over the 21st century, by implication PDI would also be expected to increase.

Emanuel's results found support from Webster et al (2005), who showed that since 1970, while the total number of hurricanes had not increased globally, the proportion of category 4 and 5 Atlantic hurricanes had doubled over this period, implying that the distribution of hurricane intensity has shifted towards being more intense.

Shepherd and Knutson (2006) argued that the actual intensity of a tropical storm mattered more for practical considerations than its potential intensity. The potential intensity was likely to be reduced by dynamical influences, local topography and other adverse factors.

An interdisciplinary team of researchers (Pielke et al 2005) surveyed the peer-reviewed literature to assess the relationships between global warming, hurricanes and hurricane impacts. They came to three broad conclusions. First, no connection has been established between greenhouse gas emissions and the observed behavior of hurricanes. The results of Emanuel (2005) were suggestive of such a connection, but they were not definitive. Second, a scientific consensus exists that any future changes in hurricane intensities will likely be small in the context of observed variability. The scientific problem of tropical cyclogenesis is so far from being solved that little can be said about possible changes in frequency. Third, under the assumptions of the IPCC, expected future damages to society of its projected changes in the behaviour of hurricanes are dwarfed by the influence of its own projections of growing wealth and population.

In order to gain a greater insight into the problem, Vecchi and Knutson (2008) attempted to analyze a much longer record of Atlantic hurricane activity spanning more than a century. If greenhouse warming causes a substantial increase in hurricane activity, then the century scale increase in global and tropical Atlantic SSTs since the late 1800s should have been accompanied by a long-term rising trend in measures of Atlantic hurricanes activity. However, the long historical tropical storm count record did not provide compelling evidence for a long term increase induced by greenhouse warming.

Knutson et al (2008) have developed a new regional modelling framework designed specifically for downscaling Atlantic hurricane activity. They used a regional climate model of the Atlantic basin which reproduces the observed rise in the number of hurricanes between 1980 and 2006, along with much of the interannual variability, when forced with observed SST and atmospheric conditions. They assessed the changes in large-scale climate that are projected to occur by the end of the 21st century by an ensemble of global climate models, and found that Atlantic hurricane and tropical storm frequencies were reduced while near-storm rainfall rates increased substantially. Their results do not support the notion of large increasing trends in either tropical storm or hurricane frequency driven by increases in atmospheric GHG concentrations.

Emanuel et al (2008) adopted another new technique for downscaling tropical cyclone climatologies from global analyses and models, which indicates that global warming should in fact reduce the global frequency of hurricanes, though their intensity may increase in some locations. This result is somewhat contrary to the earlier work of Emanuel (2005). The technique involves with the random seeding of all ocean basins with weak, warm-core vortices the great majority of which perish because of adverse environment. The surviving storms that achieve tropical cyclone strength show reasonable agreement with climatology, including ENSO-related interannual variability over the Atlantic. However, in the eastern North Pacific, the synthetic technique predicts too few events and in the north Indian Ocean too many storms are predicted. Simulated power dissipation increases over the period 1980-2006 in all basins, and is in reasonable agreement with observed power dissipation except in the eastern North Pacific. The technique is then applied to the output of seven global climate models run in support of the most recent IPCC report, Two thousand tropical cyclones in each of 5 basins were simulated using global model data from the last 20 years of the twentieth century, and the last 20 years of the twenty-second century for the A1B emission scenario. These simulations show potentially large changes in tropical cyclone activity in response to global warming, though the sign and magnitude of the changes vary a great deal from basin to basin and from model to model, reflecting large regional differences in the global model predictions as well as natural multidecadal variability. There is an overall tendency toward decreasing frequency of events in the Southern Hemisphere, but there is a tendency toward increased frequency of events in the western North Pacific. The large global increase in power dissipation over 1980-2006 cannot be ascribed to global warming, or that there is some systematic deficiency in our technique or in global models that leads to the underprediction of the response of tropical cyclones to global warming.

A very significant result of Emanuel et al (2008) is that over the north Indian Ocean, from the end of the 20th century to the end of the 22nd century, most models predict a substantial decrease in power dissipation as well as frequency. There is a divergence of model results as regards intensity and duration, and the averages are of marginal value. As far as changes in the spatial patterns of cyclogenesis are concerned, many individual models, as well as the model consensus, show increased frequency in the northern and western Arabian Sea, but decreased frequency in the Bay of Bengal.

4.6.3 Tropical Cyclones over the North Indian Ocean

Singh (2009a) has made an extensive analysis of the historical data on the frequency and intensity of cyclonic storms over the north Indian Ocean basin. There is a wide decadal variation in the number of tropical storms over the north Indian Ocean (Figure 4.6.3.1). Considering data for the months of May, October and November, which have the highest cyclonic activity, over the period 1877-1998, he has computed the pentad running frequency of severe cyclonic storms and shown that there is a clear linearly increasing trend (Figure 4.6.3.2). If the months are considered separately, the linear increasing trend is higher in November than in October and May. He also found an increasing trend in the storm intensification rate.

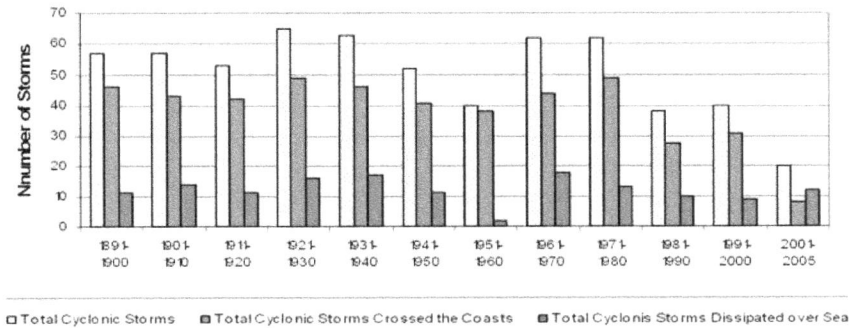

□ Total Cyclonic Storms   ◼ Total Cyclonic Storms Crossed the Coasts   ◼ Total Cyclonis Storms Dissipated over Sea

Figure 4.6.3.1 Decadal variation in the frequency of cyclonic storms over the north Indian Ocean (Source: Singh 2009a)

After the Orissa supercyclone of October 1999, there were no major storms over the Bay of Bengal or the Arabian Sea for many years. In 2007, however, there was supercyclone Gonu in the Arabian Sea in June and another intense cyclonic storm Sidr in the Bay of Bengal in November. Both these systems moved away from the Indian coastline and made landfalls in Oman and Myanmar respectively. Figures 4.6.3.3 and 4.6.3.4 show images taken by the Kalpana-1 satellite on the days when they were at their peak intensity. Figures 4.6.3.5 and 4.6.3.6 show the tracks of these two systems from their genesis to landfall.

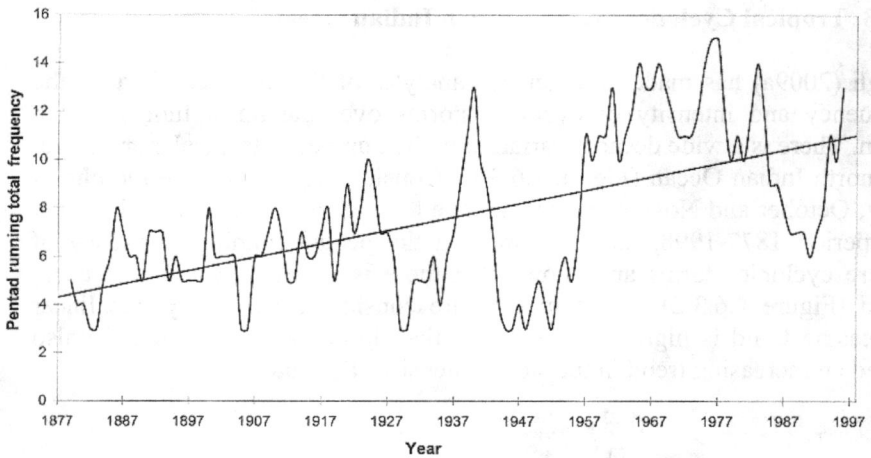

Figure 4.6.3.2 Pentad running frequency of severe cyclonic storms and linear trend over the north Indian Ocean in the months of May, October and November for the years 1877-1998 (Singh 2009a)

Table 4.6.3.1 Total number of cyclonic storms and severe cyclonic storms during the period 1891-2006 that crossed the Indian coast in different states (Source: Singh 2009a)

State	Number of cyclonic storms	Number of severe cyclonic storms
West Bengal	43	27
Orissa	78	20
Andhra Pradesh	48	34
Tamil Nadu	32	29
Kerala	0	3
Karnataka	0	2
Maharashtra -Goa	9	6
Gujarat	13	12

Figure 4.6.3.3 Kalpana-1 satellite image of supercyclone Gonu over the
Arabian Sea on 4 June 2007 (Source: IMD)

Figure 4.6.3.4 Track of supercyclone Gonu over the Arabian Sea from 1 to 7
June 2007 (Source: IMD)

Figure 4.6.3.5 Kalpana-1 satellite mage of cyclone Sidr over the Bay of
Bengal on 15 November 2007 (Source: IMD)

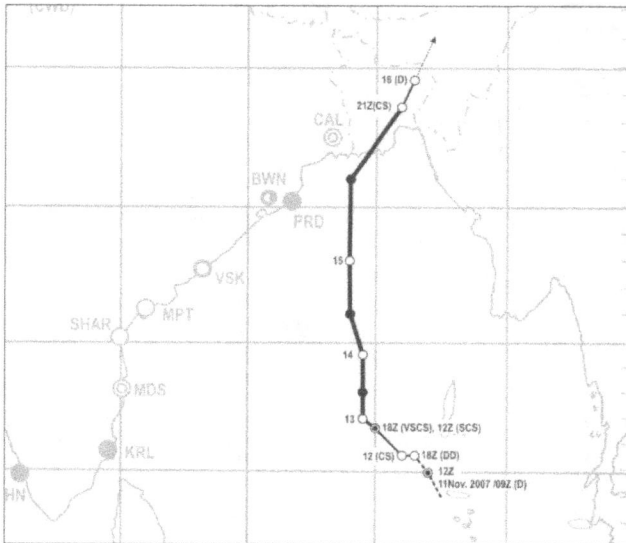

Figure 4.6.3.6 Track of cyclone Sidr over the Bay of Bengal from 11 to 16
November 2007 (Source: IMD)

On 4 June 2007, Gonu had intensified into a supercyclone or a Category 5 storm on the hurricane scale, with sustained winds of 250 km/hr, making it the most powerful cyclone ever to threaten the Arabian Peninsula on record. While tropical cyclones occasionally form in the Arabian Sea, they rarely exceed tropical storm intensity. Gonu became a tropical storm on the morning of 2 June 2007 in the east-central Arabian Sea. After some initial fluctuations in direction, it settled on a northwesterly track and began to intensify. Gonu strengthened progressively to Category 5 but then weakened to a Category 1 cyclone as it made a landfall on the Arabian Peninsula near Oman's capital city of Muscat, on 7 June 2007, but there was torrential flooding in which 28 lives were lost.

On 14 November 2007, Sidr was moving on a northward path over the Bay of Bengal as a very severe cyclonic storm or a Category 4 storm on the hurricane scale with associated winds of 210 km/hr. It made landfall over Bangladesh the next day as a weakened Category 3 storm. Over 3,000 people reportedly lost their lives, a figure which was two orders of magnitude lower than the losses in the previous Bangladesh cyclones of comparable severity, mainly because of better disaster preparedness and higher awareness among the public, and also because of a different point of landfall on the coast.

The number of cyclonic storms over the north Indian Ocean constitutes a small statistical sample, the number of severe cyclonic storms is still smaller and the instances of storms developing into supercyclones are just a few. Further, the number of landfalling systems varies widely across different sectors of the Indian coastline making some states more cyclone-prone than others (Table 4.6.3.1). Therefore, it is difficult to draw robust conclusions from a statistical analysis of historical data and derive periodicities or trends. To have two intense systems in a single year, a supercyclone in the Arabian Sea and another only marginally short of developing into a supercyclone in the Bay of Bengal, may be a climate rarity, and it cannot be said for sure that these are forerunners of more such phenomena in future. Although there is some evidence of an increasing trend, it cannot be extrapolated simplistically into the future. A lot of modeling effort is required to be put in, including factors such as projections of the sea surface temperature so that more credible and realistic conclusions can be drawn.

One such experiment has been carried out by Singh (2009b) to simulate the likely changes in the cyclogenesis patterns in the Bay of Bengal and the Arabian Sea that may result from climate change, using the regional climate model HadRM2 of the Hadley Centre of Climate Prediction and Research, U.K. In the control run, the GHG concentrations in the atmosphere were kept constant at the 1990 levels and in the simulation run, the GHGs were allowed to increase by 1% every year. Both the runs were made over the 20-year period 2041-2060. The results show that the change that is most likely to occur is a 50 % increase in the frequency of post-monsoon storms in the Bay of Bengal. However, this is likely to be offset by a decrease in the frequency of depressions and storms in the monsoon months of June-August, so that the number of storms in a year may reduce marginally, On the contrary, over the Arabian Sea, the simulations indicate a decrease of the order of 50 % in the frequency of storms in all seasons. This would mean that for the north Indian Ocean basin as a whole, the annual frequency of cyclonic storms may come down by about 15 % as a result of climate change. As regards their intensity, the simulations suggest an increase in intensity of the pre-monsoon and post-monsoon storms and a decrease in the intensity of monsoon depressions.

4.6.4 Consensus View

An International Workshop on Tropical Cyclones (IWTC-VI) was organized by the World Meteorological Organization at San Jose, Costa Rica, in November 2006. At the end of the Workshop, the participants issued a statement on the linkage between anthropogenic climate change and tropical cyclones. Since there were 125 delegates from 34 different countries and regions, and since the process was overseen by a committee of the WMO Tropical Meteorology Research Programme, the statement can be regarded as an authoritative and consensus view of the global community of tropical cyclone researchers and forecasters.

The WMO statement is remarkably balanced in its approach and findings and it is very categorical in what it says. First of all, the consensus statement makes it very clear that no individual tropical cyclone can be directly attributed to climate change. The increasing socio-economic impact that tropical cyclones have been making in recent years is largely because of rising concentrations of population and growing infrastructure in coastal regions. Another important point that the statement makes is that as of now, no firm conclusions can be drawn about the influence of global warming on tropical cyclones as there is equal evidence both for and against it.

The statement draws attention to the various difficulties in determining accurate long-term trends in the characteristics of tropical cyclones. The observed multi-decadal variability of tropical cyclones in some regions could be natural or anthropogenic or both. Methods of estimating wind speeds associated with tropical cyclone have undergone changes in recent years and different practices are followed in different regions. In most regions there are no observations from instrumented aircraft flying into tropical cyclones. The statement accepts that if the climate continues to warm, some increase in tropical cyclone peak wind speed and rainfall is likely to occur. There is, however, an inconsistency between models which project small changes in wind speed and some observational studies which suggest large changes. Also, how tropical cyclone tracks or areas of impact may change in the future cannot be foreseen now. The statement also refers to the increased vulnerability of coastal areas due to cyclone-related storm surge, if the sea level were also to rise because of global warming.

The texts of the Summary Statement on Tropical Cyclones and Climate Change and the complete statement can be referred to on the World Meteorological Organization (WMO) web site at http://www.wmo.ch/pages/prog/arep/tmrp/documents.

4.7 References

Anthwal A., Joshi V., Sharma A. and Anthwal S., 2006, "Retreat of Himalayan Glaciers – Indicator of Climate Change", *Nature and Science*, 4, 53-59.

Bindoff N. L. and coauthors, 2007, "Observations: oceanic climate change and sea level", *Climate Change 2007: The Physical Science Basis. Contribution of Working Group I to the Fourth Assessment Report of the Intergovernmental Panel on Climate Change* [Ed: Solomon S. et al], Cambridge University Press, 386-432.

Cazanave A, and coauthors, 2009, "Sea level budget over 2003-2008: A reevaluation from GRACE space gravimetry, satellite altimetry and Argo", *Global Planetary Change*, 65, 83-88.

Church J. A, and coauthors, 2004, "Estimates of the regional distribution of sea-level rise over the 1950 to 2000 period", *J. Climate*, 17, 2609-2625.

Church J. A., and White N. J., 2006: "A 20th century acceleration in global sea-level rise", *Geophysical Research Letters,* 33, L01602, doi:10.1029/2005GL024826.

Church J. A., White N. J. and Hunter J. R., 2006, "Sea-level rise at tropical Pacific and Indian Ocean islands". Global Planetary Change, 53, 155–168.

Curry J. A., Webster P. J. and Holland G. J., 2006, "Mixing politics and science in testing the hypothesis that greenhouse warming is causing a global increase in hurricane intensity", *Bulletin American Meteorological Society,* 87, 1025-1037.

Das P. K., 2001, "The climate", *The Indian Ocean - A Perspective.* Oxford IBH, New Delhi, vol.1, 55-92,

Das P. K. and Radhakrishna M. 1991, "An analysis of Indian tide-gauge records", *J. Earth System Science*, 100, 177-194.

Dasgupta S., Laplante B., Meisner C., Wheeler D. and Yan J., 2007, "The impact of sea level rise on developing countries: a comparative analysis", *World Bank Policy Research Working Paper 4136,* 51 pp.

Douglas B. C., 2001, "Sea level change in the era of the recording tide gauges" in *Sea Level Rise: History and Consequences* (Ed: Douglas B. C. et al), Academic Press, New York, 37-64.

Dvorak V. F., 1975, "Tropical cyclone intensity analysis and forecasting from satellite imagery", *Monthly Weather Review,* 103, 420-430.

Dvorak V. F., 1984, "Tropical cyclone intensity analysis using satellite data", *NOAA Technical Report NESDIS 11,* 47 pp.

Dvorak V. F. and Mogil H. M., 1992, "Tropical cyclone motion forecasting using satellite water vapor imagery", *NOAA Technical. Report, NESDIS 83.*

Emanuel K. A., 2005, "Increasing destructiveness of tropical cyclones over the past 30 years", *Nature,* 436, 686-688.

Emanuel K., 2008, "The hurricane-climate connection", *Bulletin American Meteorological Society,* 89, ES10-20.

Emanuel K., Sundararajan R. and Williams J., 2008, "Hurricanes and global warming - results from downscaling IPCC AR4 simulations", *Bulletin American Meteorological Society,* 89, 347-367.

Holgate S. J. and Woodworth P. L., 2004, "Evidence for enhanced coastal sea level rise during the 1990s", *Geophysical Research Letters,* 31, L07305, doi:10.1029/2004GL019626.

Jain S. K., 2008, "Impact of retreat of Gangotri glacier on the flow of Ganga River", *Current Science,* 95, 1012-1014.

Kireet Kumar and coauthors,. 2008, "Estimation of retreat rate of Gangotri glacier using rapid static and kinematic GPS survey", *Current Science,* 94, 258-262.

Knutson T. R. and coauthors, 2008, "Simulated reduction in Atlantic hurricane frequency under twenty-first century warming conditions", *Nature Geoscience Advance Online Publication,* doi:10.1038/ngeo202, 1-6.

Kossin J. P. and coauthors, 2007, "A globally consistent reanalysis of hurricane variability and trends", *Geophysical Research Letters,* 34, L04815, doi:10.1029/2006GL028836, 1-6.

Kulkarni A. V., Rathore B. P. and Alex S., 2004, "Monitoring of glacial mass balance in the Baspa basin using accumulation area ratio method", *Current Science,* 86,185-190.

Kulkarni A. V., Rathore B. P., Mahajan S. and Mathur P., 2005, "Alarming retreat of Parbati glacier, Beas basin, Himachal Pradesh", *Current Science,* 88, 1844-1850.

Kulkarni A. V. and coauthors, 2007, "Glacial retreat in Himalaya using Indian Remote Sensing satellite data", *Current Science,* 92, 69-74.

Landsea C. W., Harper B. A., Hoarau K. and Knaff J. A., 2006, "Can we detect trends in extreme tropical cyclones?", *Science*, 313, 452-454.

Lemke P. and coauthors, 2007, "Observations: changes in snow, ice and frozen ground", *Climate Change 2007: The Physical Science Basis. Contribution of Working Group I to the Fourth Assessment Report of the Intergovernmental Panel on Climate Change* [Ed: Solomon S. et al], Cambridge University Press, 337-383.

Le Treut H. and coauthors, 2007, "Historical Overview of Climate Change", *Climate Change 2007: The Physical Science Basis. Contribution of Working Group I to the Fourth Assessment Report of the Intergovernmental Panel on Climate Change* [Ed: Solomon S. et al], Cambridge University Press, 94-127.

Leuliette E. W, Nerem R. S. and Mitchum G. T, 2004, "Calibration of TOPEX/Poseidon and Jason altimeter data to construct a continuous record of mean sea level change", *Marine Geodesy*, 27, 79-94.

Miller L. and Douglas B. C., 2004, "Mass and volume contributions to 20th century global sea level rise", *Nature*, 428, 406-409.

MoEF 2009, "Vulnerability assessment and adaptation", *India's Initial National Communication, Chapter 3*, Ministry of Environment and Forests, Government of India, 59-132.

Pavri F., 2009, "Urban expansion and sea level rise related flood vulnerability for Mumbai (Bombay), India using remotely sensed data", *Geospatial Techniques in Urban Hazard and Disaster Analysis*, Springer Netherlands, 31-49.

Peltier W. R., 2001, "Global glacial isostatic adjustment and modern instrumental records of relative sea level history" in *Sea Level Rise: History and Consequences* (Ed: Douglas B. C. et al), Academic Press, San Diego, 65-95.

Pielke R. A. and coauthors, 2005, "Hurricanes and global warming", *Bulletin American Meteorological Society*, 86, 1571-1575.

Racoviteanu A. E., Williams M. W. and Barry R. G., 2008, "Optical remote sensing of glacier characteristics: A review with focus on the Himalaya", *Sensors*, 8, 3355-3383, DOI: 10.3390/s8053355.

Raina V. K., 2009, *Himalayan Glaciers - A State-of-Art Review of Glacial Studies, Glacial Retreat and Climate Change*, MoEF Discussion Paper, Ministry of Environment and Forests, Government of India, 55 pp.

Shankar D. and Shetye S. R., 2001, "Why is mean sea level along the Indian coast higher in the Bay of Bengal than in the Arabian Sea?", *Geophysical Research Letters*, 28, 563-565.

Shepherd J. M. and Knutson T., 2006, "The current debate on the linkage between global warming and hurricanes", *Geography Compass*, 1, 1-24.

Singh O. P., 2009a, personal communication.

Singh O. P., 2009b, "Simulations of frequency, intensity and tracks of cyclonic disturbances in the Bay of Bengal and the Arabian Sea", *Mausam*, 60, 167-174.

Singh P., Arora M. and Goel N. K., 2005, "Effect of climate change on runoff of a glacierized Himalayan basin", *Hydrological Processes*, 20, 1979-1992.

Solomon S. and coauthors (Ed), 2007, *Technical Summary, Climate Change 2007: The Physical Science Basis. Contribution of Working Group I to the Fourth Assessment Report of the Intergovernmental Panel on Climate Change,* Cambridge University Press, 91 pp.

Thayyen R. J., 2008, "Lower recession rate of Gangotri glacier during 1971-2004", *Current Science*, 95, 9-10.

Thompson B., Gnanaseelan C., Parekh A, and Salvekar P. S., 2008, "North Indian Ocean warming and sea level rise in an OGCM", *J. Earth System Science*. 117, 169-178.

Trenberth K., 2005: "Uncertainty in hurricanes and global warming", *Science,* 308, 1753-1754.

Trenberth K. E. and coauthors, 2007, "Observations: surface and atmospheric climate change", *Climate Change 2007: The Physical Science Basis. Contribution of Working Group I to the Fourth Assessment Report of the Intergovernmental Panel on Climate Change* [Ed: Solomon S. et al], Cambridge University Press, 236-336.

Unnikrishnan A. S., Rupa Kumar K., Fernandes S. E., Michael G. S. and Patwardhan S. K., 2006, "Sea level changes along the Indian coast: observations and projections", *Current Science*, 90, 362-368.

Unnikrishnan A. S. and Shankar D., 2007, "Are sea-level-rise trends along the coasts of the north Indian Ocean consistent with global estimates?", *Global Planetary Change*, 57, 301-307.

Wang C. and Lee S.-K., 2008. "Global warming and United States landfalling hurricanes", *Geophysical Research Letters*, 35, L02708, doi:10.1029/2007GL032396. 1-4.

Webster P. J., Holland G. J., Curry J. A. and Chang H.-R., 2005, "Changes in tropical cyclone number, duration, and intensity in a warming environment", *Science*, 309, 1844-1846.

WMO, 2009, "WMO statement on the status of the global climate in 2008", *WMO No. 1039,* World Meteorological Organization, Geneva, 13 pp.

Chapter 5

Climate Change Impacts on Monsoon, Agriculture and Human Health

Monsoons are observed over many parts of the world, in Asia, Africa, Australia and America, but the Indian southwest monsoon stands out amongst all of them. It is the strongest of all monsoons, it has linkages with the global atmospheric circulation, and it is an important component of the earth's total climate system. The Indian southwest monsoon is India's only source of water. It sustains the livelihood of millions of Indian farmers and influences food production. It is a dominant factor in shaping India's economic growth rate. It has moulded Indian culture and tradition, inspired poets, and set the notes of Indian classical music. The Indian southwest monsoon is indeed "*the* monsoon" (Kelkar 2009).

The monsoon is beautiful and captivating, but at times it brings misery and death. It is a regular visitor, but may turn up earlier or later than expected. It makes promises, but does not always keep them. The monsoon rainfall is grossly uneven and India has some of the wettest places on earth and also the driest. The rainfall is not uniform in time either, being interspersed with dry spells. Each year's monsoon is a unique blend of cloud and sunshine and in the strictest sense, it has no past analogues.

5.1 Monsoon Rainfall

There are two reassuring facts about the Indian southwest monsoon. One is that it comes every year regularly without fail, although the nature of the onset may be different from one year to another, sometimes weak and delayed, or sometimes early and strong as in 2009 (Figure 5.1.1). The other is that the All-India Summer Monsoon Rainfall (AISMR), averaged for the country as a whole and over the period 1 June to 30 September, has not shown any increasing or decreasing trend over the last 135 years (Figure 5.1.2.3).

Figure 5.1.1 Onset of southwest monsoon over Kerala on 23 May 2009, nine days ahead of its normal date of 1 June as seen in this Kalpana-1 satellite image (Source: IMD)

5.1.1 Global Warming and the Monsoon

Ever since the issue of global warming came to the forefront, the instant reaction of every Indian has been to ask whether the monsoon will be affected and how. The impending threat of global warming to the Indian monsoon is like an attack on the lifeline of the nation. Even when the prediction of the monsoon a season in advance can go wrong, as was the case in 2009, predicting the state of the monsoon 50 to 100 years from now is a daunting problem. There are indications (Mani et al 2009) that global warming may reduce the predictability of the monsoon even on the shorter time scales. Nevertheless, a credible climate scale prediction of the monsoon is an utmost necessity and a high priority concerted effort is required to be made in that direction. We must know for sure where the monsoon is heading, so that we can plan our strategies for ensuring water and food security for our population while maintaining the rate of economic development.

From a meteorological point of view, the monsoon is essentially an annual oscillation of the state of the atmosphere in response to the relative position of the sun, as it moves between the tropic of Cancer in the northern hemispheric summer and the tropic of Capricorn in the southern hemispheric summer. However, in the larger sense, the monsoon is an extremely complex phenomenon that involves not only the atmosphere, but land and ocean as well. It has attracted the curiosity of scientists from around the world, and of course India, but their understanding of the monsoon is yet far from complete, and the phenomenon is such that it has eluded even a precise and unique definition.

The British scientist, Halley (1686), put forth the hypothesis that the Indian monsoon was caused by the differential heating between the Asian landmass and the Indian Ocean. In other words, the monsoon has the character of a giant land-sea breeze that reverses its direction twice during a year. In April, when the sun starts heating the land, the southwest monsoon begins and blows until October; then the land cools and the northeast monsoon blows in the winter until April again. This was the first ever scientific explanation of the Indian monsoon. Halley's three-century old hypothesis does not explain many things about the monsoon that we know today and it has many modern critics. However, it goes to Halley's credit that land surface temperatures over the Eurasian continent and sea surface temperatures over the Indian Ocean are the two factors that have continued to dominate all efforts to understand and predict the monsoon, though of course in an increasingly complex manner.

An interesting and often overlooked aspect of global warming is that it is not uniform all across the world. The warming trends observed in the past as well as the temperatures projected for the future under different scenarios show large regional differences. Figure 5.1.1.1 shows the anomalies of annual average surface temperature (°C) for the period 1850-2008 relative to the average for 1961-1990, separately for the northern and southern hemispheres as obtained from the HadCRUT series. Dashed circles have been overlaid upon the graphs to highlight the different rates of warming over the two hemispheres in recent years. The warming rate is much higher over the land-covered northern hemisphere compared to the predominantly oceanic southern hemisphere. This means that the Eurasian continent is warming at a faster rate than the Indian Ocean, providing an increased land-sea thermal contrast. From very simplistic considerations this can be interpreted as being favourable for the strengthening of the Indian summer monsoon in future.

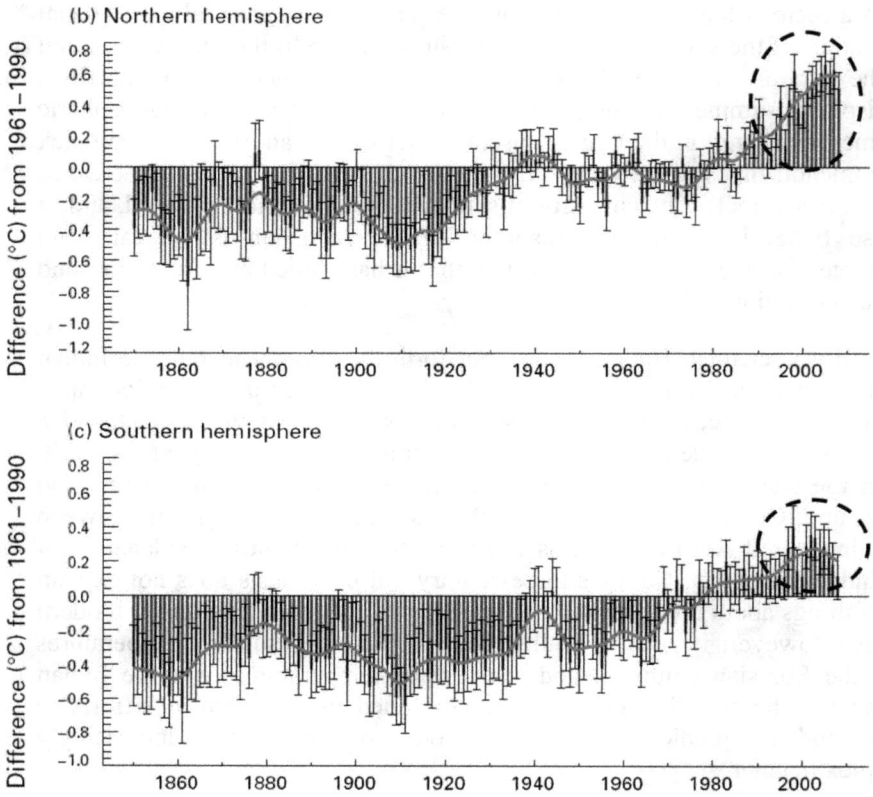

Figure 5.1.1.1 Anomalies of annual average northern and southern hemispheric near-surface temperature (°C) for the period 1850-2008 relative to the average for 1961-1990 Dashed circles are overlaid to highlight the different rates of warming over the two hemispheres in recent years (Source: WMO 2008)

5.1.2 Temperature and Precipitation Trends over India

In the particular context of global warming and climate change, it is of great importance to investigate whether temperature and rainfall over India have been showing any signs of an increasing or decreasing trend. Of course even if a trend exists, it does not mean that one can just linearly extrapolate it indefinitely into the future.

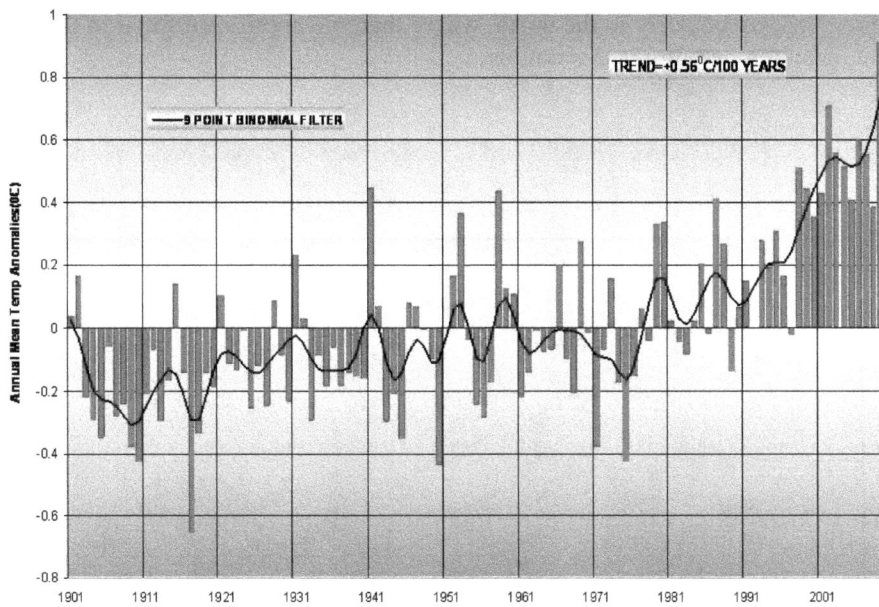

Figure 5.1.2.1 Annual mean surface temperature anomalies (°C) over India
for the period 1901-2009 relative to the average for 1961-1990.
(Source: IMD)

Figure 5.1.2.1 shows the all-India annual mean temperature anomalies for the period 1901-2009 with respect to the 1961-1990 mean which has a value of 24.64 °C. According to IMD's latest estimates, the annual mean temperature averaged over India as a whole in 2009 was +0.913 °C above the 1961-1990 average. The year 2009 was thus the warmest year on record since 1901. The other warmer years on record in order are 2002 (0.708), 2006 (0.6), 2003 (0.560), 2007 (0.553), 2004 (0.515), 1998 (0.514), 1941 (0.448), 1999 (0.445), 1958 (0.435), 2001 (0.429), 1987 (0.413) and 2005 (0.410). It is worth noting here that the ranking of warm years over India has been done on consideration of the third decimal point or one-thousands of a degree which should be well within the error bars associated with the temperature values.

Over the period 1901-2009 which is longer than a century, the annual mean temperature has exhibited an increasing trend of +0.56 °C/100 yr. However, this value does not apply all over the country. If the spatial pattern of annual temperature anomalies is examined (Figure 5.1.2.2), it is seen that there are some pockets of Rajasthan, Gujarat and Bihar where the trend has been -0.56 °C/100 yr indicating that the temperatures are in fact falling. There are also some regions, particularly in the north, where there no significant trend in the annual mean temperature is discernible.

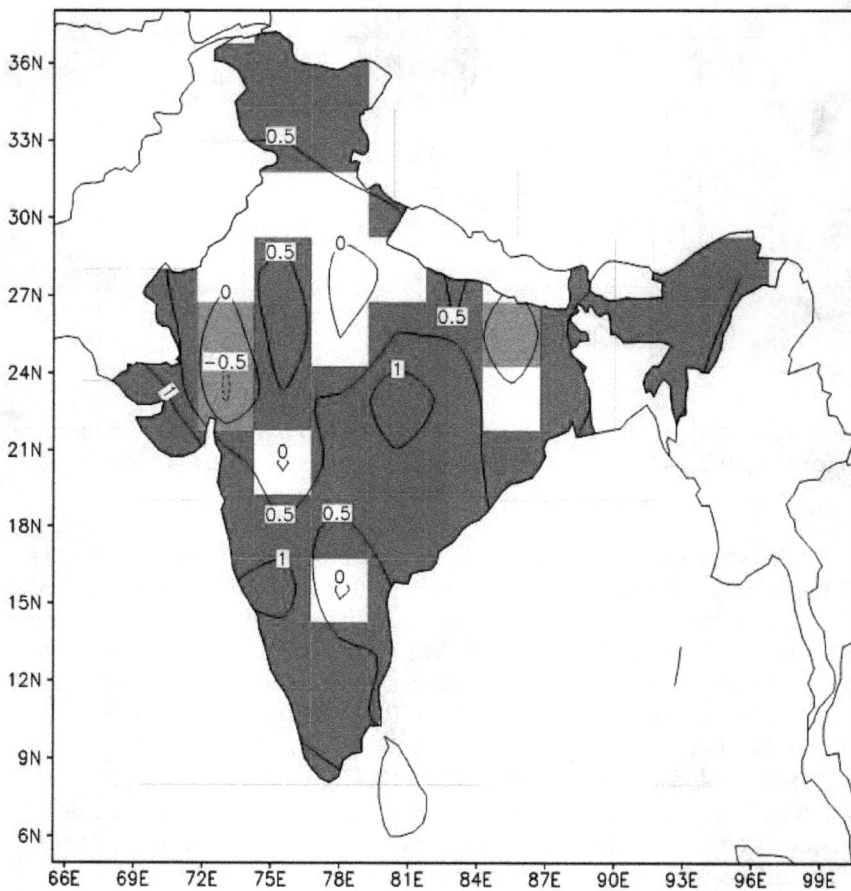

Figure 5.1.2.2 Trends in annual mean surface temperature over India (°C/100 yr) for the period 1901-2009 (Source: IMD)

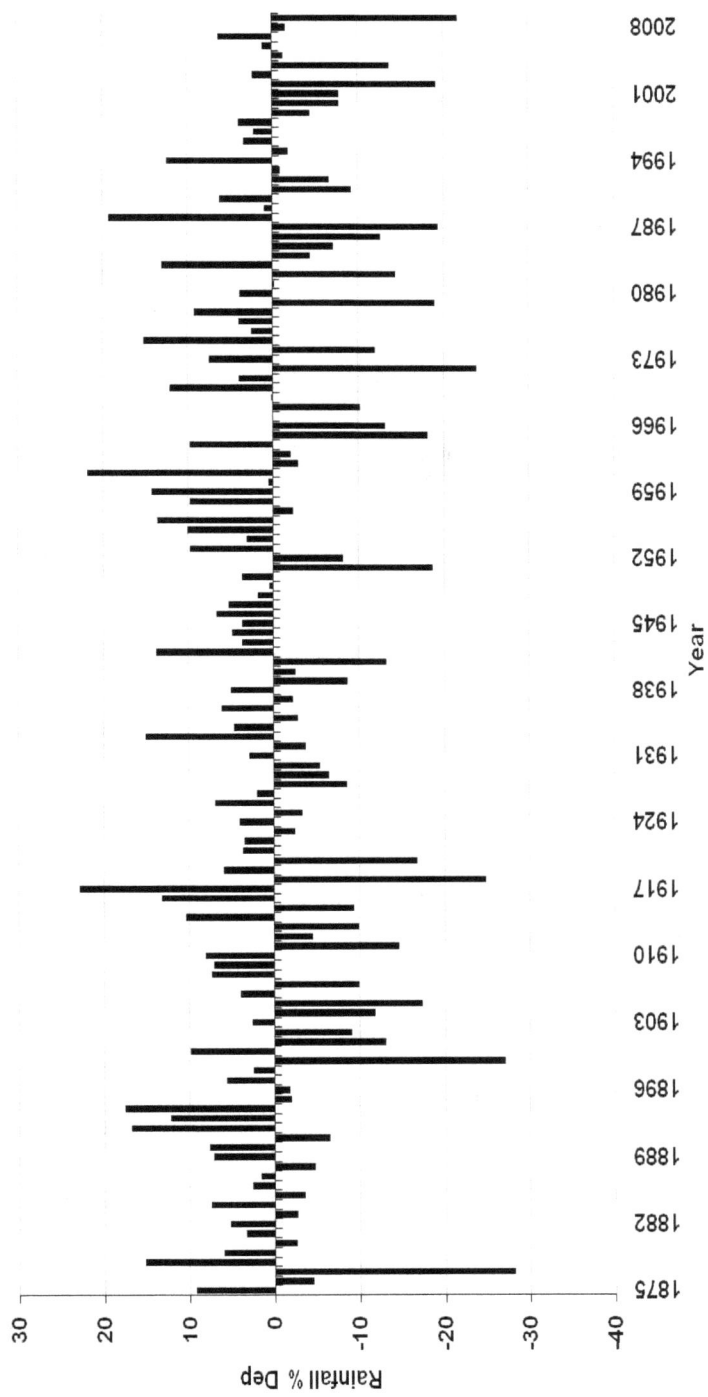

Figure 5.1.2.3 Interannual variability of the AISMR (in percentage departure from normal) over the years 1875-2009 (Source: IMD)

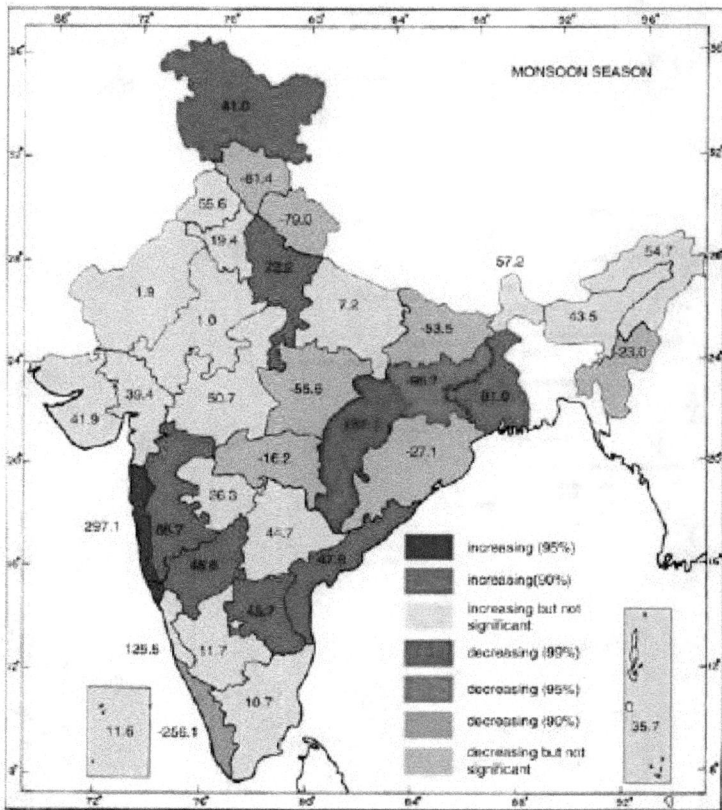

Figure 5.1.2.4 Increase/decrease in rainfall (mm/100 years) over the meteorological subdivisions of India for the southwest monsoon season. Different levels of significance are shaded (Source: Guhathakurta et al 2006)

Guhathakurta et al (2006) have subjected the revised All-India Summer Monsoon Rainfall (AISMR) data series for the 103-year period 1901-2003 to a low-pass filter in order to suppress the high frequency oscillations. The weights used were 9-point Gaussian probability curves (0.01, 0.05, 0.12, 0.20, 0.24, 0.20, 0.12, 0.05 and 0.01). They found no evidence of any linear trend in this series. Even when they carried out a similar exercise for the rainfall of each of the monsoon months June to September separately and applied linear regression technique and the Student's t-test, they did not find any trend. In other words, there is no long-term trend in the seasonal as well as the monthly monsoon rainfall over India. This is an important conclusion, because many times fears are expressed that the pattern of monsoon rainfall, including its distribution within the season, is changing.

Guhathakurta et al (2006) also carried out a linear trend analysis to examine the long term trends in rainfall over different meteorological subdivisions and the monthly contribution of each of the monsoon months to annual rainfall. During the southwest monsoon season, 3 subdivisions viz., Jharkhand, Chattisgarh and Kerala, showed significant decreasing trend and 8 subdivisions viz., Gangetic West Bengal, West Uttar Pradesh, Jammu and Kashmir, Konkan and Goa, Madhya Maharashtra, Rayalaseema, Coastal Andhra Pradesh and North Interior Karnataka, showed significant increasing trends (Figure 5.1.2.4). The remaining 25 subdivisions did not exhibit any trend, either increasing or decreasing, in the monsoon rainfall.

5.1.3 Climate Projections of Monsoon Rainfall

In the IPCC AR4 (Christiansen et al 2007), regional climate projections of area-averaged temperature and precipitation changes have been presented from the coordinated set of climate model simulations called the Multi-Model Data set (MMD) using the SRES A1B scenario through the 21st century. All of Asia is very likely to warm during this century, especially in the interior of the continent. Over south Asia, summer precipitation is likely to increase, and an increase in the frequency of intense precipitation events is very likely, but monsoonal flows and the tropical large-scale circulation are likely to be weakened. Monsoon processes show up quite differently in different climate models and there is uncertainty in quantifying estimates of projected precipitation changes. A paradox of the results of climate simulations is that while monsoonal rainfall is projected to increase, the monsoon circulation is projected to weaken.

The MMD models generally exhibit a cold bias as well as a dry bias when compared with the observed climate. They capture the general regional features of the monsoon, but most of them have a problem in simulating the observed heavy monsoon rainfall over the west coast of India, the north Bay of Bengal and northeast India. This is mainly due to their coarse resolution which tends to smooth out out the orographic processes that produce the heavy rainfall. Most of the MMD-A1B models project a decrease in precipitation over south Asia in the winter months of December-February and an increase during the monsoon months June-August (Figure 5.1.3.1). The projected pattern of temperature rise is shown in Figure 5.1.3.2.

The MMD-A1B models show a median warming of 3.3 °C by the end of the 21st century. The median change in precipitation is 11 % by the end of the 21st century, and seasonally it is -5 % in December-February and 11% in June-August, with a large inter-model spread. Only 3 of the 21 models project a decrease in annual precipitation.

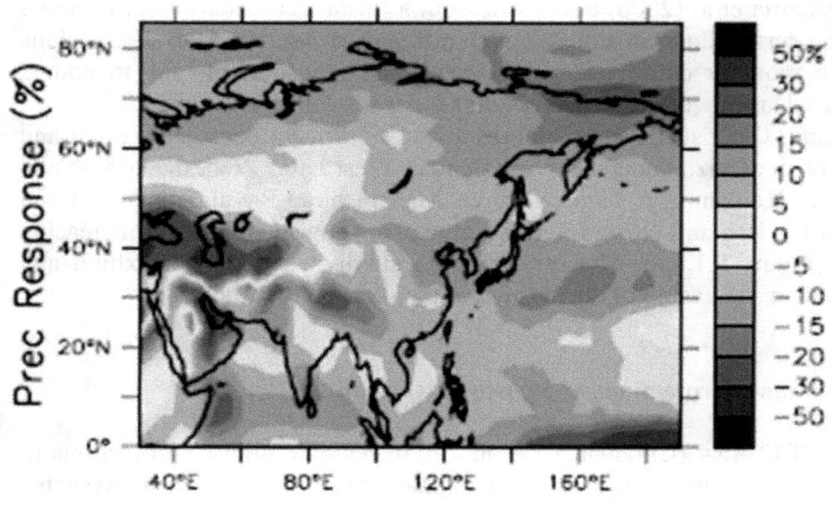

Figure 5.1.3.1 Projections of precipitation change (%) from 1980-1999 to 2080-2099 over Asia for June-July-August from the MMD A1B simulations averaged over 21 models. (Source: IPCC AR4, Christensen et al 2007)

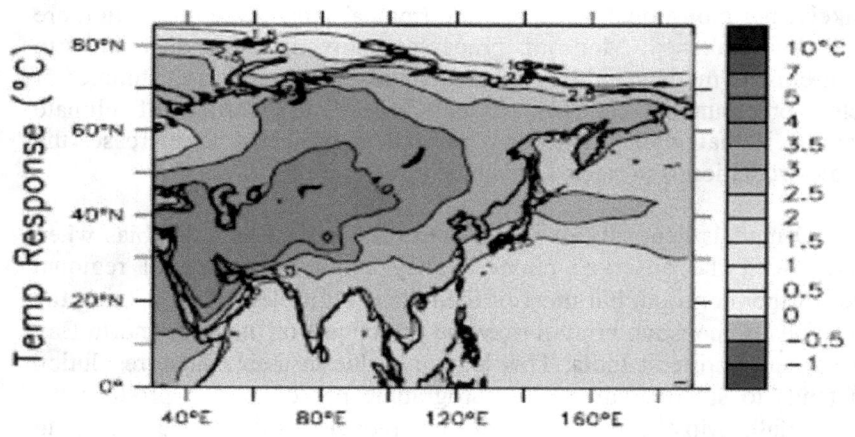

Figure 5.1.3.2 IPCC projections of temperature change (°C) from 1980-1999 to 2080-2099 over Asia for June-July-August from the MMD A1B simulations averaged over 21 models. (Source: IPCC AR4, Christensen et al 2007)

Rupa Kumar et al (2006) have carried out regional climate model simulations for India, based on the second generation Hadley Centre regional climate model. This model is known as PRECIS (Providing Regional Climates for Impacts Studies) and it is a limited area high- resolution atmospheric and land surface model that can be located over any part of the globe (Jones et al 2004). The boundary conditions are derived from a high-resolution AGCM (HadAM3H) with a horizontal resolution of 150 km, in the so-called time slice experiments. Instead of running coupled models for century-long integrations, two time slices, namely 1961-1990 and 2071-2100, were selected from 240-year (1860-2100) long transient simulations with HadCM3. Ensembles of three baseline simulations for the period 1961-1990, three simulations for the A2 future scenario (2071-2100) and one simulation for the B2 future scenario (2071-2100) have been run with HadAM3H and assessed. PRECIS has been configured for a domain extending from about $1.5° N$ to $38° N$ and $56° E$ to $103° E$ with a resolution of about 50 km.

PRECIS simulations under scenarios of increasing greenhouse gas concentrations and sulphate aerosols indicate a marked increase in surface air temperature in the 2071-2100 time frame as compared to the 1961-1990 baseline for both A2 and B2 scenarios, but the B2 scenario shows slightly less increase. The warming is expected throughout India, but there could be substantial spatial differences. Extremes in maximum and minimum temperatures are also expected to increase into the future, but with the night temperatures increasing more than the day temperatures.

In the PRECIS model simulations of Rupa Kumar et al (2006) described above, the mean monsoon rainfall for the baseline 1961-1990 was 939 mm with a standard deviation of 57 mm. Thus it has overestimated the mean rainfall while it has underestimated its variability. There were several other differences between the simulated and observed rainfall, particularly in quantitative terms. The mean simulated JJAS rainfall for 2071-2100 is 1114 mm for the A2 scenario and 1078 mm for the B2 scenario which translates into a 20% increase in the rainfall compared to 1961-1990.

The maximum increase in rainfall of the order of 40 % is seen over western Maharashtra and northeast India for both A2 and B2 scenarios. The increase is about 10-30 % over most of central and eastern India and there is a decrease in rainfall of 5-10 % over extreme northwest India. In other words, the projection is an accentuation of the present spatial rainfall variability over India with the wet regions getting wetter and the dry regions getting drier.

In a study by Tanaka et al (2005), intensities and trends of Hadley, Walker, and monsoon circulations were compared for the IPCC 20th Century simulations (20C3M) and for 21st century simulations (SRES A1B scenario), using the upper tropospheric (200 hPa) velocity potential data. In response to a global warming scenario, it is anticipated that the Hadley circulation may become weaker by 9 %, Walker circulation by 8 %, and monsoon circulation by 14 % by the late 21st century as an ensemble mean of the IPCC model simulations. However, such results are to be viewed against the background of the poor capability of the current AOGCMs in reproducing and predicting the tropical circulations. While a weakening of the monsoon circulation would suggest a reduction in monsoon rainfall, there are other model projections of an increased monsoon precipitation in a global warming scenario because of increased availability of total precipitable water in spite of a weaker circulation.

There are some studies (Kitoh and Uchiyama 2006) of the onset and withdrawal times of the Asian summer rainfall season in 15 MMD simulations. The results are indicative of a delayed withdrawal of the monsoon rains, while the onset dates are largely unaffected. This again amounts to an increase in the total monsoon rainfall. The results, however, are not clear and not very consistent among models.

The response of the Asian summer monsoon to a possible doubling of CO_2 concentration has been the subject of many investigations in previous years However, these simulations were based upon single models chosen by the investigators and the results varied widely. Some studies showed that the monsoon was not likely to be affected at all, others projected a decline in monsoon precipitation, and many indicated that the monsoon rainfall could increase.

A concerted attempt in this domain of investigation has been that of Kripalani et al (2007), who have analysed the behaviour of 22 coupled climate models in the IPCC AR4 data base and their response to a doubling of CO_2. The study of Kripalani et al does not offer a complete evaluation of the performance of each of the models or their relative merits, but is limited to the simulation of the total Indian southwest monsoon seasonal precipitation and its interannual variability (Tables 5.1.3.1 and 5.1.3.2). Out of the 22 chosen models, seven models simulated an annual cycle that was similar to the observed one in terms of shape and magnitude and six other models simulated the shape of the annual cycle well but underestimated the precipitation amounts, particularly during the spring and the summer periods. Another group of six models showed the rainfall maximum occurring a month later than observed, resulting in the underestimation of rainfall during spring and summer. The JJAS precipitation total varied between 50 and 91 cm across these 19 models while the coefficient of variation ranged from 3 to 13 %. The remaining three models were unable to simulate the annual cycle to the desired degree of accuracy. Thus there is no single coupled climate model which can be regarded as ideal from the point of view of the Indian monsoon.

Table 5.1.3.1 Climate models participating in the IPCC AR4 experiments (Source: Kripalani et al 2007)

No.	Originating group	Country	IPCC ID
1	Bjerknes Centre for Climate Research	Norway	BCCR-BCM2.0
2	Beijing Climate Center	China	BCC-CM1
3	National Centre for Atmospheric Research	USA	CCSM3
4	Canadian Centre for Climate Modelling and Analysis	Canada	CGCM3.1
5	Meteo-France/Centre National de Recherches Meteorologiques	France	CNRM-CM3
6	CSIRO Atmospheric Research	Australia	CSIRO-Mk3.0
7	Max Planck Institute for Meteorology	Germany	ECHAM5/MPI-OM
8	Meteorological Institute of University of Bonn/ METRI of KMA	Germany/ Korea	ECHO-G
9	LASG/Institute of Atmospheric Physics	China	FGOALS-g1.0
10	NOAA/Geophysical Fluid Dynamics Laboratory	USA	GFDL-CM2.0
11	NOAA/Geophysical Fluid Dynamics Laboratory	USA	GFDL-CM2.1
12	NASAGoddard Institute for Space Studies	USA	GISS-AOM
13	NASA/Goddard Institute for Space Studies	USA	GISS-EH
14	NASA/Goddard Institute for Space Studies	USA	GISS-ER
15	Institute for Numerical Mathematics	Russia	INM-CM3.0
16	Institut Pierre Simon Laplace	France	IPSL-CM4
17	Center for Climate System Research (The University of Tokyo)/National Institute for Environmental Studies and Frontier Research Center for Global Change (JAMSTEC)	Japan	MIROC3.2 hires
18	Center for Climate System Research (The University of Tokyo)/National Institute for Environmental Studies and Frontier Research Center for Global Change (JAMSTEC)	Japan	MIROC3.2 medres
19	Meteorological Research Institute	Japan	MRI-CGCM2.3.2
20	National Center for Atmospheric Research	USA	PCM
21	Hadley Centre for Climate Prediction and Research/ Meteorological Office	UK	UKMO-HadCM3
22	Hadley Centre for Climate Prediction and Research/Meteorological Office	UK	UKMO-HadGEM1

Table 5.1.3.2 Simulation of observed precipitation and projection for double CO_2 by various climate models (Source: Kripalani et al 2007)

IPCC ID of model	Simulation of observed precipitation features			JJAS precipitation amount (cm)	
	Shape of annual precipitation cycle	Monthly precipitation amounts	Month of peak precipitation	Simulation of observd amount	Projection for 2x CO_2
BCCR-BCM2.0	Well-simulated	Well-simulated	Well-simulated	71	
BCC-CM1	Incorrect	Incorrect	Incorrect	-	
CCSM3	Well-simulated	Well-simulated	Well-simulated	78	
CGCM3.1	Well-simulated	Well-simulated	Well-simulated	80	87 (+8.5%)
CNRM-CM3	Well-simulated	Well-simulated	Well-simulated	78	83 (+5.4%)
CSIRO-Mk3.0	Not properly simulated	Underestimated in spring/summer	One month later than observed	50	
ECHAM5/MPI-OM	Well-simulated	Well simulated	Well-simulated	74	77 (+2.9%)
ECHO-G	Well-simulated	Underestimated	Well-simulated	60	
FGOALS-g1.0	Not properly simulated	Underestimated in spring/summer	One month later than observed	62	
GFDL-CM2.0	Not properly simulated	Underestimated in spring/summer	One month later than observed	61	
GFDL-CM2.1	Well-simulated	Well-simulated	Well-simulated	76	
GISS-AOM	Not properly simulated	Underestimated in spring/summer	One month later than observed	57	
GISS-EH	Well-simulated	Underestimated	Well-simulated	63	
GISS-ER	Well-simulated	Underestimated	Well-simulated	61	
INM-CM3.0	Well-simulated	Underestimated	Well-simulated	63	
IPSL-CM4	Incorrect	Incorrect	Incorrect	-	
MIROC3.2 hires	Not properly simulated	Underestimated in spring/summer	One month later than observed	91	97 (+6.9%)
MIROC3.2 medres	Well-simulated	Well-simulated	Well-simulated	90	97 (+8.2%)
MRI-CGCM 2.3.2	Well-simulated	Underestimated	Well-simulated	60	
PCM	Incorrect	Incorrect	Two peaks	-	
UKMO-HadCM3	Not properly simulated	Underestimated in spring/summer	One month later than observed	72	84 (16.6%)
UKMO-HadGEM1	Well-simulated	Underestimated	Well-simulated	58	

Six models were chosen for projection experiments with a 1 % increase in CO_2 per year until the amount is doubled. This is about twice the rate of increase in CO_2 due to anthropogenic factors. The projected and simulated annual cycles for the monsoon derived from all six models show an increase in summer precipitation and perhaps an extension of the monsoon period. This may signify an intensification of the monsoon system. It is interesting to note that the simulated multi-model ensemble annual cycle accurately resembles that of the observed. The projected ensemble annual cycle clearly reveals an increase in precipitation during the summer and the following period but its simulation of observed precipitation is an underestimate. The projected increase in monsoon precipitation has a wide difference across the six models, ranging from 2.9 % to 16.6 %. The ensemble projects an 8% increase of JJAS precipitation from 81 to 87 cm.

Detailed inferences about the impact of global warming on the monsoon can however, be drawn only with the climate models. Initial research work on the simulation of the circulation features and rainfall patterns of the Indian monsoon was mostly done in an isolated manner with individual atmospheric general circulation models of varying design and capabilities. These studies have more often than not, resulted in generating diverse and conflicting views of the future behaviour of the monsoon and its variability. It is only recently that a large number of atmosphere-ocean coupled general circulation models have been run in a coordinated way to assess their performance and suitability for monsoon prediction on a climate scale. In a qualitative sense, most of the climate models seem to agree on one point that the monsoon precipitation is likely to increase across the 21st century in a global warming scenario with increasing CO_2. Thus the present consensus view emerging out of the climate model runs is that of an intensification of the monsoon, and there is a possibility that is not indicated by all models, that the monsoon season may extend somewhat longer than its current period.

However, the real difficulty arises when it comes to a quantification of the results. Most climate models are unable to simulate the observed features of the Indian monsoon in their totality. Many models underestimate the monsoon rainfall while some of them cannot simulate the observed monthwise precipitation pattern or the peak precipitation month. Hence only a few models can be trusted with the job of making a century scale prediction and even these have yielded diverse results (Table 5.1.3.1). There is no climate model currently available internationally that can be truly relied upon from all angles pertaining to the monsoon. The increase in precipitation resulting from a doubling of CO_2 in the atmosphere is likely to be anywhere from 2.9 to 16.6 % for the A1B scenario (Table 5.1.3.2). This large uncertainty is further compounded by the possibility of the doubling of CO_2

not occurring in reality or the global CO_2 emissions not conforming to the A1B scenario.

In arriving at climate scale projections of the monsoon, a basic prerequisite is that the climate model used must be able to reproduce closely the known climatology of monsoon precipitation, otherwise the projected pattern of rainfall cannot be regarded as trustworthy. The main difficulty arises here because of the inadequacies of convective parameterization scheme which lead to systematic biases in the model.

Mukhopadhyay et al (2009) have investigated the impact of different convective closures on systematic biases of Indian monsoon precipitation climatology in a high-resolution regional climate model. For this purpose, they have run the Weather Research Forecast (WRF) model at 45 and 15 km (2-way nested) resolution with three different convective parameterization schemes, Grell-Devenyi (GD), Betts-Miller-Janjic (BMJ) and Kain-Fritsch (KF) for the period 1 May to 31 October of the years 2001 to 2007. The model was forced with the NCEP/NCAR (National Center for Environmental Prediction / National Center for Atmospheric Research) reanalysis as initial and boundary conditions. The simulated monsoon rainfall patterns for the months June-September with the three different convective schemes were then compared with that observed.

Mukhopadhyay et al found that out of the three schemes which they applied, only BMJ was able to produce a reasonable mean monsoon rainfall pattern. BMJ and KF underestimated the lighter rain rates and overestimated the heavier ones while GD did the opposite. GD systematically overestimated the lighter rain rate and underestimated the moderate rain rate throughout the season, whereas BMJ and KF had problems in the initial stages only. Overall, KF had a moist bias and GD had a dry bias. Such studies only highlight how difficult the problem is, and there is no straightforward or unique solution. They serve as a reminder to us that when we look at climate projections of monsoon rainfall, we should remember the biases of the models.

In another recent study, Preethi et al (2009) have compared the performance of seven global fully coupled ocean-atmosphere models in simulating Indian summer monsoon climatology as well as its interannual variability. They have used multi-member 1-month lead hindcasts made by several European climate groups as part of the DMETER programme (Development of a European Multi-model Ensemble system for seasonal to inter-annual prediction). The dependence of the model simulated Indian summer monsoon rainfall and global sea surface temperatures on model formulation and initial conditions were studied in detail. They found a large spread among the ensemble members in the simulations of the monsoon rainfall indicating that the initial conditions were influencing the results.

In summary, all indicators such as past statistical trends in AISMR, inter-hemispheric differences in the rate of global warming and projections of climate models, point to a strengthening of the Indian southwest monsoon along with the march of the 21st century, but there are large uncertainties when it comes to quantifying the likely change in monsoon rainfall.

It is a discouraging fact that there has been no effort on the part of Indian scientists to build an Indian model in spite of the continuous upgradation of computational resources and the growth of institutional infrastructure in India. The approach has all along been to test foreign models over Indian conditions and judge their usefulness for predicting the monsoon. The search for an ideal model that could be used for monsoon prediction on the seasonal and climate scale is still on.

5.2 Extreme Events

One of the manifestations of climate change is an increase in the intensity of extreme weather events and in the frequency of their occurrence. Even where temperature or rainfall have not shown a perceptible change in the average value, it is possible that their range of variation around the mean value may have increased. So if there is a steady decrease in the minimum temperature and it is accompanied by an increase in the maximum temperature, the mean daily temperature would remain unchanged but the change would show up through an increased number of cold waves and heat waves. Similarly, the mean daily precipitation amount may not show a change, but there could be an increase in the duration and intensity of both dry spells and wet spells.

According to the IPCC AR4 (Solomon et al 2007), in the last 50 years for the land areas sampled, there has been a significant decrease in the annual occurrence of cold nights and a significant increase in the annual occurrence of warm nights. The distributions of minimum and maximum temperatures have not only shifted to higher values, consistent with overall warming, but the cold extremes have warmed more than the warm extremes over the last 50 years. More warm extremes imply an increased frequency of heat waves. A prominent evidence of a change in rainfall extremes is the increase in the number of heavy precipitation events over the mid-latitudes in the last 50 years, even in places where mean precipitation amounts are not increasing.

Mani et al (2009) have found that over the central Indian region, convective instability is increasing and also becoming increasingly easier to realize.

They found a statistically significant increasing trend in the convective available potential energy (CAPE) over the region during the June–September monsoon season and a clear decreasing trend in the convective inhibition energy (CINE). A rise in the frequency of extreme events which are the result of convective instability of the atmosphere, is therefore to be expected.

In India, the occurrence of extreme events has been a topic of interest and the focus of media attention in recent years. Every time there is an extreme weather event like very heavy rainfall or a heat wave or a long break in the monsoon, the commonly asked question is whether it is a forerunner of more such events likely to happen in future.

By definition, an extreme weather event is one which does not commonly occur at a given place and in a given season, and it is extreme only in a relative sense. A temperature of 45 ºC on a summer afternoon in Jaisalmer would not be categorised as an extreme event, but it would certainly be regarded as one if Shimla recorded that temperature. When long period climate normals are computed, extreme values get averaged out, losing the attention they deserve. It is only when an extreme event assumes the nature of a disaster, with heavy loss of life and property, that it becomes a matter of importance, examination, discussion and even controversy.

The Indian monsoon comes with a reassuring regularity, but it exhibits a wide range of variability on the spatial, temporal, intraseasonal, interannual and decadal scale. This makes all the difference between floods and droughts, between Cherrapunji and Jaisalmer, between Mumbai and Pune. When the monsoon rains are timely and equitable, we do not bother, but when they are not, reality dawns once again. The monsoon has always had its vagaries and it is going to show them in future too. That is why every rainfall related extreme event over India should not be regarded as an indicator of climate change (Kelkar 2005).

On the time scale of a day, there are many past instances of Indian stations having recorded as much as half of their annual rainfall, or even more than their annual rainfall, in one single day (Dhar et al 1981). Rainfall of 50 cm or more in a 24-hour period is not an uncommon phenomenon at all (Rakhecha et al 1980).

A case in point is what has come to be known as the Mumbai rain event of 26 July 2005. The observatory at Santa Cruz in north Mumbai recorded a rainfall of 94.4 cm during the 24 hours ending at 8:30 am on 27 July 2005, while the Colaba observatory in Mumbai's southern tip recorded barely 7.3 cm in the same period. Rainfall over Vihar lake was 105 cm, even higher than Santa Cruz. The previous record of heaviest 24-hour rainfall over Mumbai was 58 cm for Santa Cruz and 37 cm for Colaba on 5 July 1974. Comparatively speaking, only Santa Cruz broke the previous record, but for Colaba the rainfall was in no way unusual.

On 5 July 1974, Mumbai had received 58 cm in a single monsoon day and the city had taken it in its stride. On 13 July 2000, Mumbai had recorded exceptionally heavy rains: Vasai 49, Thane 45, Santa Cruz 37 and Colaba 25 cm. However, what happened in Mumbai on 26 July 2005 was truly unprecedented. Never before perhaps had the metropolis experienced anything like it. Suburban trains normally running at intervals of 3 minutes, came to a grinding halt and 150,000 commuters including schoolchildren got instantly stranded at railway stations. Buses were unable to ply and the roads were bursting to capacity with stagnant northbound traffic. Telephone lines, mobile phone services and power supply broke down in many areas. Hundreds of people lost their lives because of drowning or by getting trapped in cars. Highways connecting the city got blocked and the airport had to be closed. The island city was really marooned.

Many different reasons have been ascribed to the Mumbai flooding of 26 July 2005. The two main causes cited are the uncontrolled urbanisation of north Mumbai and the destruction of mangroves and the inadequacy of the existing drainage system. It is evident that in the process of housing construction and setting up of industries, the waterway that allows the accumulated rain water to drain out has been drastically reduced. Large slum colonies have encroached upon the storm water drains and the Mithi river, which is Mumbai's main river.

River floods are common during the monsoon season. They can be predicted in advance because there is considerable time lag between the occurrence of heavy rainfall in the upper catchment and the consequent build-up of the flood flow in the river, and its travel to a downstream area. Such a lead time is not available in case of drainage congestion caused by local rainfall. For this reason, urban flooding has become a matter of concern for India. In today's world, there can be no argument against development. No one can be denied the right to a better living. But urban growth has to be controlled and planned on the strong foundations of foresight and discipline. What happened in Mumbai is waiting to happen in many other urban areas in India, where drainage systems have become obsolete and inadequate, and river beds have been encroached upon by illegal housing colonies. Climate change has very little to be blamed for extreme events like urban flooding.

5.3 Agriculture

There is strong evidence that India had to suffer from recurrent famines throughout its history, and all the more so during the period of British rule. Famines frequently affected many large regions and in some years even the entire country, bringing misery and death to millions of people. Although famines were primarily caused by a failure of the crops due to extremely poor monsoon rains, the difficulties of the population got compounded by the general apathy of the rulers, lack of relief provisions, or inconsiderate measures like increased taxation being introduced at the same time.

The infamous Bengal famine of 1770 was the first to have occurred under the regime of the British East India Company, and is said to have resulted in the starvation death of a population of ten million in that province. During the nineteenth century, famines kept on affecting different parts of the country, and millions of people died of hunger, since they had no alternative sources of livelihood. Bengal was struck by famine again in 1866 and at the same time Orissa was also badly affected. However, this calamity also had a positive outcome. The commission of enquiry appointed by the British government to examine the circumstances of the famine recommended the setting up of an all-India meteorological organization. This far-reaching recommendation, which was supported by other bodies like the Asiatic Society of Bengal, eventually led to the establishment in 1875 of the India Meteorological Department (IMD 1975).

In 1876, there was a famine in Madras, and there was a countrywide famine in 1899. The year 1918 is remembered as the year of the great Indian famine. In 1943, there was yet another major famine in Bengal leading to the death of at least three million people. India's British rulers were at that time preoccupied with the Second World War, and left the Indian farmers to fend for themselves. Whatever they had harvested was also acquired by the government in the name of the war effort, and grain trading was banned.

Even after India won its struggle for independence in 1947, it had to carry on for another twenty years with its battle on the food front. With the partition of the country, it had lost some of its most fertile land and the population had been increasing continuously. In the decade of the 1960s, India's imports of food grain had shot up to over 10 million tonnes per year. But then came the Green Revolution and it transformed the country into a self-sufficient nation. The building of huge buffer stocks, a good public distribution system, and an efficient relief and disaster management organisation, have all freed the Indian population from the miseries of famine. The statistics in Table 5.3.1 speak for themselves.

Today, India has a total arable land of 162 million hectares. Out of this, the irrigated area is 52 million hectares, which is the largest amongst all the countries of the world. India is currently the world's largest producer of tea, milk, pulses and jute, the world's second largest producer of wheat, rice, groundnut, vegetables, fruits and sugarcane, and the third largest producer of potatoes and cotton. Since independence, the Indian agricultural sector has made significant strides in all directions, and the per capita availability of food grains in India has increased considerably in spite of the ever-growing needs of its massive population. However, the monsoons still continue to play a dominating role in Indian agriculture. The meagre rainfall over many regions and topographical features set a limit to the percentage of land that can be irrigated. Furthermore, irrigation water does not have any independent source, it is again the monsoon. Excepting the Himalayan rivers which are fed by snowmelt, all other Indian rivers originate out of the monsoon rain that falls into their catchment areas. The monsoon can thus be said to be India's source of water for agricultural and all other purposes.

Table 5.3.1 India's food grain production, imports, buffer stock and population 1950-2000
(Data Source: Good News India web site http://goodnewsindia.com)

	1950	1960	1970	1980	1990	2000
Food grain production (million tonnes)	50.8	82.0	108.4	129.6	176.4	201.8
Food grain import (million tonnes)	4.8	10.4	7.5	0.8	0.3	-
Buffer stock (million tonnes)	-	2.0	-	15.5	20.8	40.0
Population (million)	361	439	548	683	846	1000

The peculiarity of India's annual rainfall is that about 75-80 % of it occurs in the monsoon months of June to September. Maximum use of the rain water has to be made in these four months and water stored for use in the remaining eight months of the year. Indian agriculture has traditionally got adjusted with this rainfall pattern. The kharif crop is the rainfed crop that is raised directly on the monsoon rains or with some supplemental irrigation if available, and the rabi crop is raised on the residual soil moisture that the monsoon has left behind after its withdrawal in October. In north India, the occasional winter rains help to boost the crop productivity, particularly in the case of wheat, but elsewhere only those crops that can do without water can be grown in the rabi season. For southern India, especially Tamil Nadu, however, the main agricultural season is associated with the northeast monsoon months of October to December.

The percentage of the irrigated area to the total agricultural land area in India varies widely from state to state. Only Punjab and Haryana are in a fortunate position with more than 80-90 % land under irrigation while Uttar Pradesh, Bihar and West Bengal have a half or more of irrigated land. For most other states, the figures are around 10-30 %. As a result, for the country as a whole, only about 45 % area has availability of irrigation water (Figure 5.3.1). The remaining 55 % area is completely dependent upon rainfall and it contributes to about a half of the total food grain production. The irrigation potential has largely been achieved over many parts of the country and there is not much scope to bring more land under irrigation in the future.

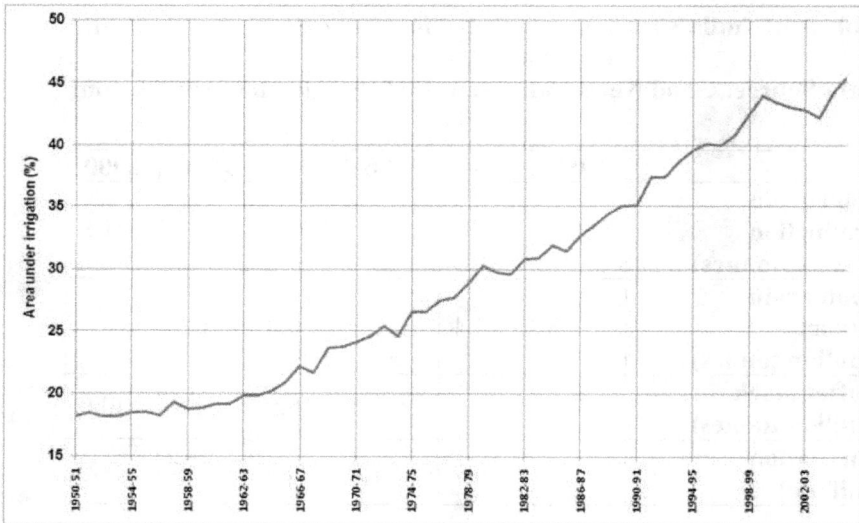

Figure 5.3.1 Growth of percentage area under irrigation in India
(Source: Department of Agriculture and Cooperation, Ministry of Agriculture, Government of India web site
http:// dacnet.nic.in/eands/At_Glance_2008)

Although the broad pattern of Indian agriculture is tuned to the annual distribution of rainfall, it is the vagaries of the monsoon and the distribution of monsoon rainfall across the country and within the season, that heavily impact agricultural production. The timeliness of monsoon onset, the timing of active and break phases, the duration of dry and wet spells, the effects of weather on the incidence of pests and diseases, all influence the food grain production that is achieved at the end of the kharif season. The cropping pattern over the country has evolved in consonance with the long term climate prevailing in various parts of the country but the crop acreage and production in any given year depends upon the amount and distribution of rainfall that was actually available on the smaller spatial scales vis-à-vis the water requirements of specific crops at their critical growth stages. Thus drought can occur at the district or subdivision level even when the country as a whole has received a statistically normal rainfall.

The influence of the monsoon rains on the country's agriculture is clearly evident in the graph of the country's food grain production over the 40-year period from 1966-67 to 2007-08 (Figure 5.3.2). It shows a steady rising trend that is attributable to the continuing improvements in technology, farming practices, increase in area sown and availability of irrigation. In fact the food grain production has more than doubled over the last 40 years. However, the graph has peaks that correspond to the exceptionally good monsoons of 1970, 1975 and 1983, and dips that correspond to the all-India droughts of 1966, 1972, 1979, 1982, 1987 and 2002. The worst monsoon failure in recent years occurred in 2009 when the all-India rainfall deficit was 22 %. The official crop production figures are not yet available, but preliminary estimates place the 2009 kharif food grain output at 99 million tones which would be 22 million tonnes below the record kharif production in 2007 but still better than for the previous drought year of 2002. Another noticeable feature of this graph is the rise in rabi production to the extent that it is now almost catching up with the kharif season crops.

Besides directly impacting the agricultural production of the country, large scale deficiencies of monsoon rainfall cause many other problems such as shortage of fodder for animals and scarcity of drinking water, while excess rains lead to flooding, disruption of normal life and loss of standing crops. During the monsoon season, crops are also lost to weather-related pests and diseases.

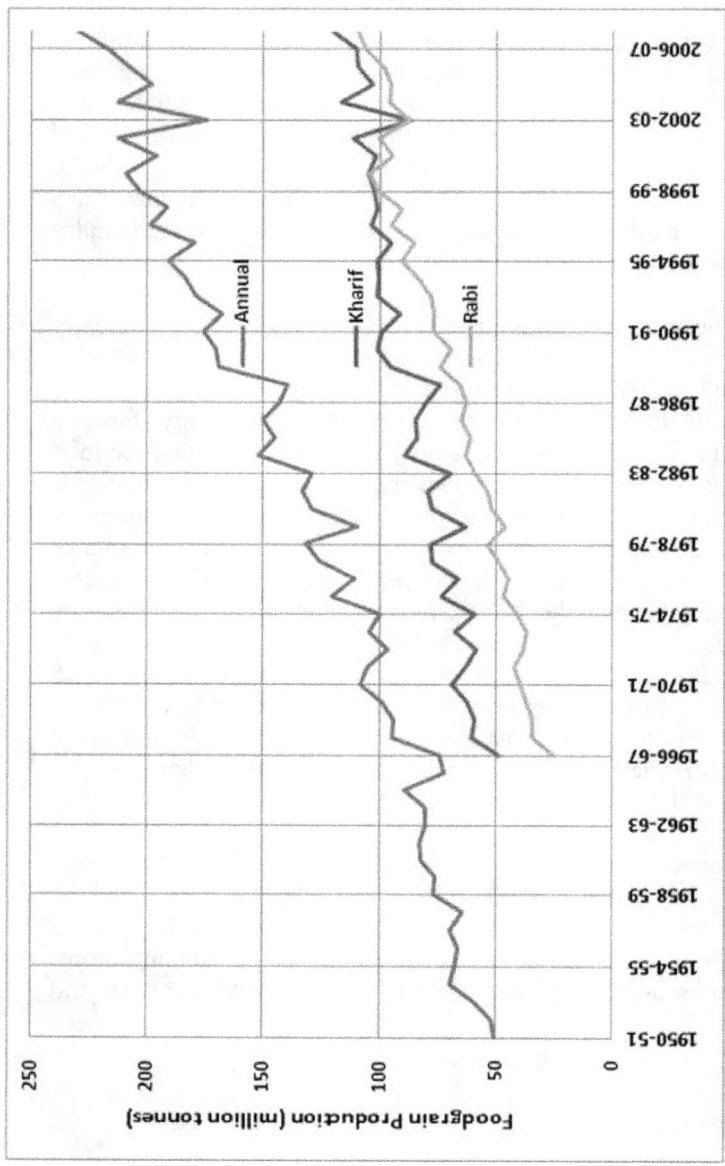

Figure 5.3.2 Food grain production in India during kharif and rabi seasons and annual total, from 1950-51 to 2007-08. (Source: Department of Agriculture and Cooperation, Ministry of Agriculture, Government of India web site http:// dacnet.nic.in/eands/At Glance 2008)

5.3.1 Climate Change and Indian Agriculture

While the possible impact of climate change needs to be quantified as realistically as possible so that mitigation actions can be initiated, effects of short term climate variability on Indian agriculture should also be investigated as they are indicators of what could happen in future. For example, India raised a bumper wheat crop of 75.5 million tonnes in 1999-2000, which exceeded the previous year's production by 5 million tonnes. This was not due to the application of any new technology, but because of anomalously low temperatures that prevailed in the months of January-March 2000 and gave more time to the wheat crop for grain formation and filling. The opposite situation prevailed in 2004 when temperatures over the Indo-Gangetic plains remained 3-6 °C above normal in the month of March, causing the wheat crop to mature a fortnight earlier and lowering the wheat production by 4 million tonnes compared to the previous year. Variations in crop productivity brought about by short term climate variability may in fact be even higher in the case of crops like rice, pulses or oilseeds, which are largely rainfed.

When considering the impact of climate change on agriculture, it must be kept in mind that while climate changes have also occurred in the past, they have taken place gradually over thousands of years, giving ecosystems the time needed to adapt themselves to the changing climate. In comparison, the climate change that they are now being subjected to is occurring very rapidly, making the ecosystems much more vulnerable. There is a need to assess (a) the sensitivity of the ecosystem, or how much it will respond to a given change in climate, (b) its adaptability or how far it could adjust to the change or anticipated change, and (c) its vulnerability or how much it is likely to be affected if it is sensitive to climate change but cannot adapt to it. Generally speaking, arid lands, semi-arid lands, low-lying coastal areas, floodprone areas and small islands are going to be more vulnerable to climate change.

Climate change is likely to affect Indian agriculture both directly and indirectly. Crop productivity can be directly affected by changes in carbon dioxide concentration, air temperature and moisture, availability of solar radiation and rainfall, and indirectly through changes in soil conditions and the incidence of crop pests and diseases. The effects of long term changes to temperature, CO_2 concentration and rainfall on crop growth and productivity can actually be measured by growing crops in controlled environments like greenhouses or phytotrons, However, it is both difficult and expensive to replicate such physical experiments in all the agroclimatic regions and with a wide variety of crops. Crops respond differently to changes in levels of atmospheric CO_2 and they have different requirements of water, sunshine and temperature. Pests and diseases of crops thrive in different weather conditions. Crop response may also depend significantly on management practices and strategies adopted. The climate change impacts on Indian agriculture can therefore be assessed more conveniently and rigorously with the help of crop growth models in which changes in environmental parameters, individually as well as in different combinations, can be simulated and the results analyzed critically.

Some of the popularly used crop models are the Wheat Growth Simulator (WTGROWS), the ORYZA models for rice, the Decision Support System for Agro-technology Transfer (DSSAT) for different crops and InfoCrop, which is an indigenously developed decision support system (Aggarwal et al 2006, Kalra et al 2007, Krishnan et al 2007). The InfoCrop modelling framework requires limited inputs and also includes databases of typical Indian soils, weather and genotypes. The current version of the model deals with a wide range of crops like chickpea, cotton, groundnut, maize, mustard, pearl millet, pigeonpea, potato, rice, sorghum, soybean, sugarcane and wheat.

However, results of investigations of the likely impacts of climate change on Indian agriculture have a wide spread depending upon the complexity of the model chosen and the inputs provided to the model. Complex models do not necessarily come up with accurate results, as the type and volume of data that they demand as input are many times not available, forcing approximations to be made. Many investigations have been made on what would happen to crops if the present CO_2 concentration of 385 ppm was increased to a value of 550 ppm which it is likely to reach by the year 2050. Under this scenario, the yields of some crops such as rice, wheat, legumes and oilseeds may go up by about 10-20 %. However, a 1 °C increase in temperature may reduce the yields of wheat, soybean, mustard, groundnut, and potato by 3-7 %. Therefore, the effects of increasing CO_2 may be offset by the effects of increasing temperature to some extent. The productivity of most crops is estimated to decrease marginally by the year 2020 but by as much as 10- 40 % by 2100. The increased incidence of extreme events like droughts, floods and heat waves will cause higher production variability and losses. The length of the growing period of crops in rainfed areas is also likely to reduce, especially in peninsular regions and southern India.

Ortiz et al (2008) have found that some wheat growing regions of the world may benefit from the rise in temperature due to global warming, but in some areas where the temperatures are already in the optimal range, a further increase in temperature may result in reduced productivity. For the Indo-Gangetic Plains which currently account for 15 % of global wheat production, they infer that by 2050 as much as a half of its area might have to be reclassified as a heat-stressed, irrigated, short-season production environment. Some other Indian studies estimate a possible loss of 4-5 million tonnes in India's wheat production with every rise of 0.1 °C temperature even after considering carbon fertilization.

5.3.2 Problems in Crop Response Modelling

Crop growth models vary in their degree of complexity depending upon how the plant, soil and atmospheric processes and the interactions between them are incorporated in the model. While interpreting the results of such impact models it is important to be aware of their complexity and how near they are to simulating real situations in the control runs. For example, crop yields will increase with higher availability of sunshine as well as with increase in rainfall, However, sunshine and rainfall are themselves are inversely related. Hence very simple crop response models are likely to produce contradictory or exaggerated results (Figure 5.3.2.1).

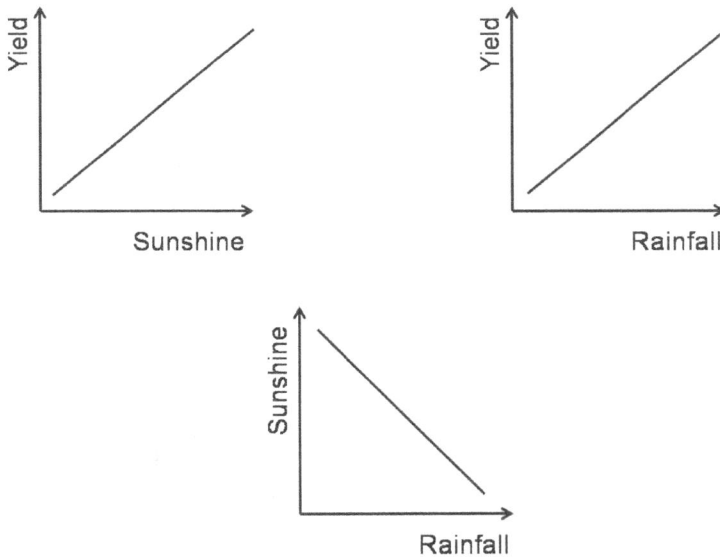

Figure 5.3.2.1 Schematic crop response models showing positive relationships of crop yield with sunshine and rainfall while there is a negative relationship between sunshine and rainfall

The impact of elevated CO_2 on agriculture is not uniform over all crops. Generally speaking, a rise in the CO_2 level leads to an increase in the photosynthesis rate, leaf area index through better light interception, water use efficiency and biomass and crop yield. Increased CO_2 levels therefore lead to higher crop productivity but the effect tapers off after a significant initial rise. The actual response depends upon the crop species and soil fertility. The process of photosynthesis gets enhanced in several crops as atmospheric CO_2 increases. Such carbon fertilization effects are important for C3 crops such as rice, wheat, soyabean and pulses. The yields of these crops in general, are expected to increase by 10-15% as CO_2 goes up from 370 ppm to 550 ppm. On the other hand the C4 crops like maize and sugarcane have a low response to CO_2.

The general methodology used for assessing the impact of climate change on agriculture is to first assess the likely extent of future climate change by running a climate model assuming one or more appropriate scenarios and then obtain the crop response to it by using a crop growth simulation model. A good model is one which incorporates all the effects of increased CO_2, such as the enhancement of photosynthesis, increased plant growth, reduction in stomatal opening, reduction in transpiration per unit leaf area and increase in water use efficiency. The test of a good model lies in the proper simulation of all phenological phases from sowing or transplantation to maturity. There are however several problems in crop modelling. For example, in many models, the impact of global warming is studied by simply raising the maximum and minimum temperature baselines as a whole, which is not likely to happen in reality. Likewise, the rainfall anomaly that could result from climate change is applied as a whole across the entire growing season. This again is a simplistic assumption as the number of rainy days and the distribution of rain over the season may not change proportionally. In some cases, the parameters are modelled independently of each other and their interactions get ignored. Some thresholds used in the models may be very critical and even if they are changed by a small amount or are in error, the results may get significantly altered.

The genetic coefficients used in many crop response models may not be applicable directly to local crop varieties and may need to be calibrated. A major uncertainty in crop modelling comes from the indirect effects of temperature and humidity conditions and persisting cloudiness on the incidence of pests and diseases. Whether climate change can lead to the development of environmental conditions that are favourable to the growth of certain pests or diseases, needs to be investigated independently as they can seriously affect crop growth and yields. Similarly, the effects of altered soil conditions, salinity and erosion and the adoption of new management practices have to be factored into the impact analysis.

Another source of uncertainty comes from the climate model chosen for projecting the likely future changes in temperature, rainfall, CO_2 and other parameters which form the input to the crop models. The climate model that is used may have a coarse resolution and it may not be suitable for agricultural applications. The climate scenarios chosen may also be inappropriate from the agricultural point of view.

InfoCrop is a generic dynamic crop model (Aggarwal et al 2006), which has been indigenously developed and overcomes many of the difficulties mentioned above. It provides an integrated assessment of the effect of weather, differences in crop variety, pests, soil and management practices on crop growth and yield. It also takes into account soil nitrogen and organic carbon dynamics in aerobic as well as anaerobic conditions, and greenhouse gas emissions..

5.3.3 Adaptation Potential

While pessimistic scenarios can be painted about the adverse effects of climate change on Indian agriculture, they have a silver lining because it is possible to initiate adaptation measures that have the potential to counteract many negative impacts. In fact, adaptation is a natural and continuing process in agriculture and diversity is one such manifestation of climatic adaptation in nature. For counteracting climate change it is only necessary to enhance the present adaptive capacity of agriculture.

Although Indian farmers have accumulated their traditional knowledge over the ages, have applied it to their advantage, they have always been open to the induction of new technological inputs and practices that are economically viable. In fact the success of India's green revolution was largely due to the flexibility that the farmers showed towards trying out new crop varieties and farming practices. The traditional coping strategies are mixed cropping, changing land use, diversification of income sources and migration. However, the nature of Indian agriculture has undergone a significant change in recent years. While in the past it was subsistence-oriented, it is now increasingly becoming market-oriented. There is therefore a need for new adaptation strategies to cope with the problems associated with climate variability and climate change.

Whether new adaptive measures would be acceptable to Indian farmers would depend of many factors such as their cost, the availability of appropriate financial systems to enable farmers to bear the cost, and easy access to new technology. Farmers have to be encouraged to overcome farm level constraints, and to introduce new soil and water management practices. On the national scale, there would be a need to overcome biophysical constraints like adverse soil characteristics and topography by breeding new varieties of crops, introducing new cropping patterns and developing innovative schemes and concepts like agro-forestry.

Ramakrishna et al (2009) have emphasized the need for preparation of contingency plans for effective management of the impacts of climate change on agricultural production levels. They recommend the strengthening of research in the area of agricultural meteorology and development of integrated climate crop modelling systems on different scales suitable for decision making. It is necessary to build good data bases and bring in remote sensing and GIS techniques for better delineation of regions vulnerable to climatic change and extreme weather events.

5.3.4 Methane and Rice Cultivation

Methane (CH_4) is a greenhouse gas that is a very minor atmospheric constituent compared to CO_2 but it is important because its warming potential is 21 times greater than that of CO_2. The global concentration of methane in the atmosphere has continuously risen since 1850 and it is now more than 150 times what it was then. Methane has been recognized in the Kyoto Protocol as a potent contributor to global warming. The sources of atmospheric methane are both natural and anthropogenic. The highest contribution to atmospheric methane concentration comes from wetlands, while the ocean and fresh water bodies, gas hydrates and termites also produce methane. Processes such as oxidation by chemical reaction with tropospheric hydroxyl (OH), stratospheric oxidation and microbial uptake by soils constitute methane sinks.

Among the anthropogenic sources of global methane emissions, livestock, rice cultivation, natural gas and oil production, coal mining and biomass burning are some of the major contributors. In 1994 India's total methane emission was 18,000 Gg out of which 78% came from the agriculture sector (MoEF 2009a), Enteric fermentation contributed 9000 Gg, rice cultivation 4000 Gg and biomass burning 1600 Gg. Therefore, in the context of climate change and Indian agriculture, methane assumes a peripheral but important role. India is currently the second largest producer of rice in the world, and as India's population increases in the coming years, rice cultivation would have to be increased for meeting the food requirements of the growing population. How much of this would contribute to global warming is an issue that needs serious consideration. Methane campaigns in India have shown that extensive and authentic methane emission measurements are needed for coming to a realistic conclusion and that results from isolated experimental plots should not be just extrapolated on a national scale. It is also necessary to chalk out and adopt mitigation strategies very carefully. Moreover, sooner or later, international pressure is going to build up against India to reduce its national methane emission. This would force India to rethink both on its production and consumption of rice. Reducing the production of rice is an agricultural matter but weaning people away from consuming rice is a social issue that would involve changing eating habits of millions of Indians.

5.4 Human Health

Weather has an important relationship with the well-being of people, both emotionally and physically. Getting up to see overcast skies can lead to depression and a miserable day, while a bright sunlit morning can make one feel cheerful for the rest of the day. Raindrops and snowflakes, and the early showers of the monsoon are associated with romance. On the physiological side, the rainy season is the time when it is easy to catch a cold or develop cough and fever, Heat strokes are common in summer, while winter is the time for the skin to become dry and itchy. Pre-monsoon haze and dust can cause breathing problems, while a rise in humidity in the monsoon season can aggravate ailments like arthritis. Many times doctors recommend patients a change of weather as a therapeutic measure. There is no doubt that human health is closely related to to weather and climate. A possible change in the climate is therefore regarded as a threat to human health.

Environmental and atmospheric conditions vary widely across different land regions of the world, depending upon latitude, altitude, distance from the coastline, and such other factors. Human beings, however, have the capability in the long run to adapt themselves physiologically to the climate in which they live and to evolve compatible lifestyles. That is why human habitations have been able to exist and even thrive in the most adverse climates ranging from deserts to cold icy continents to rain forests. The human body can adjust itself well to diurnal and seasonal variations which are regular in nature, and it also has some resilience to face temporary deviations from the normal climate. However, severe weather aberrations and extreme events can result in morbidity as well as mortality. This explains why people living in Rajasthan, for example, can take summer temperatures persisting at 48 °C in their stride, but an heat wave lasting for only a few days can take a high toll of human lives in states like Andhra Pradesh or Orissa. In 2003, there was a heat wave over Europe, in which thousands of people died, particularly older people who had lacked mobility and were confined to homes that had no cooling facilities. The impact of climate change on human health is therefore most likely to be experienced through the increased frequency of extreme events related to temperature.

There is another way in which climate change is likely to impact health and that is through extreme events of heavy rainfall. Such events can lead to localized as well as large scale flooding. The stagnant water and wet soil become a breeding ground for mosquitoes which are responsible for spreading malaria. The quality of drinking water also gets affected by flooding, and water-borne diseases like cholera and diarrhea begin to spread among the population. In 1994, western India had seen intense heat waves during the summer and temperatures as high as 50 °C had been recorded in some parts, which allowed disease-carrying vectors to breed. Later during the monsoon months it so happened that the same areas experienced heavy flooding, with the result that a malaria epidemic broke out over Gujarat. Another instance was the Orissa supercyclone of October 1999, which was followed by a malaria epidemic that severely affected the local population.

In India, malaria is the most common vector-borne disease. In the year 2007, as many as 1.48 million cases of malaria alone had been reported in the country and there were 1173 reported deaths on its account. Other major vector-borne diseases in India are chikungunya, kala azar, filariasis, dengue and Japanese encephalitis but their prevalence is far less compared to malaria. Their number of cases reported annually is only in thousands for the entire country and the mortality rate is also very low (Dhiman 2009). The concern before India is to know beforehand how these disease vectors will react to climate change particularly in terms of rising temperature and increased flooding due to heavy rainfall events.

In the health sector, India's greatest concern with respect to climate change is about the possible resurgence of malaria. The National Malaria Eradication Programme launched in the early 1950s had been highly successful and it had resulted by 1965 in a spectacular reduction in the morbidity and mortality rates associated with malaria. However, malaria returned subsequently in a big way, peaking in 1976, as mosquitoes developed resistance to pesticides and the parasites became immune to anti-malarial drugs. The government thereafter had to modify its malaria containment strategies and launch new national programmes for malaria control.

Temperatures in the range of 20-30 °C are optimal for the development and transmission of most of the malaria vectors, and relative humidity higher than 55 % is optimal for their longevity and successful completion of sporogony. The transmission windows are therefore different over different parts of India, the date of arrival of the monsoon being another important consideration. Besides these weather related parameters, other factors such as urbanization, availability of irrigation water, and the level of socioeconomic development also play a role in the transmission dynamics.

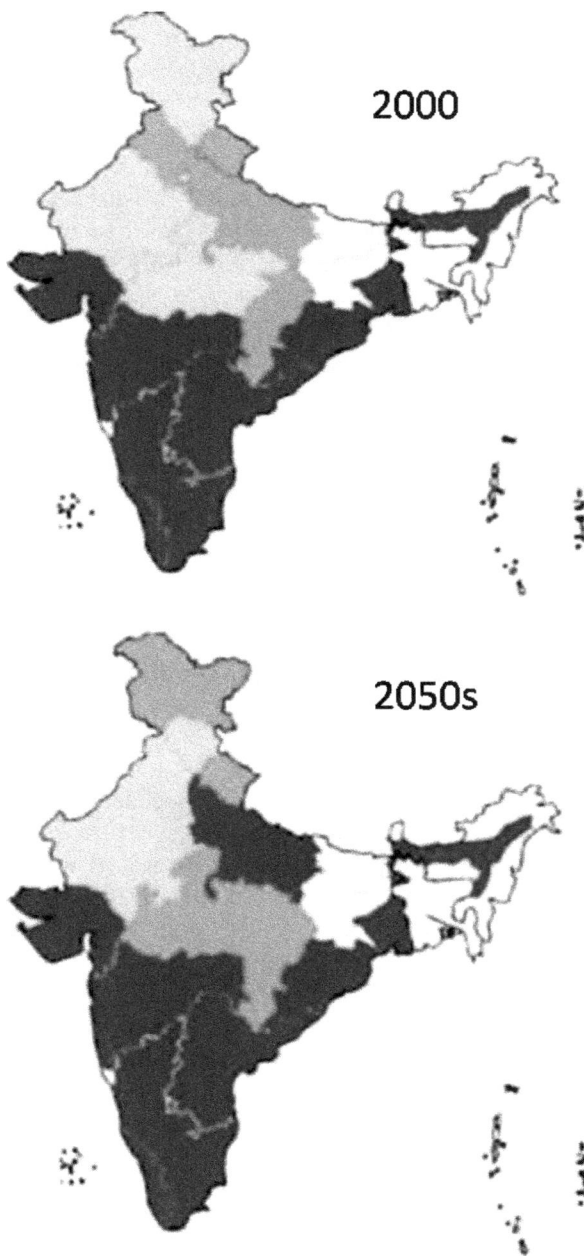

Figure 5.4.1 Malaria transmission window in different states of India, for base year 2000 and projection for 2050s under climate change scenario. Colour code: brown 10-12, orange 7-9, yellow 4-6 months in a year. (Source: MoEF 2009b)

Peri-urban malaria is a newly emerging type of malaria introduced into urban areas by migrants having chronic malaria and living in poor environmental conditions in makeshift dwellings that favour breeding and transmission.

The malaria transmission window calculated on the basis on temperature and humidity considerations alone, remains open for 10 to 12 months in a year in the states of Tamil Nadu, Kerala, Karnataka, Andhra Pradesh, Maharashtra, Gujarat, Orissa, West Bengal and Assam, for 7 to 9 months in a year in Chhattisgarh, Uttar Pradesh, Uttarakhand, Punjab and Haryana, and for less than 6 months in other states. Considering a 3.8 °C increase in temperature and a 7 % increase in relative humidity by the 2050s with reference to the present, the length of the transmission window will increase in the states of Madhya Pradesh, Uttar Pradesh and Jammu and Kashmir (Bhattacharya et al 2006, MoEF 2009b). Thus the southern states such of Karnataka, Kerala, Tamil.Nadu and Andhra Pradesh appear to be less vulnerable to climate change whereas the northern states such as Jammu and Kashmir, Himachal Pradesh, Punjab, Haryana, Uttarakhand, Uttar Pradesh and the northeastern states may be more vulnerable to climate change (Figure 5.4.1).

Such general projections based entirely upon climate considerations, however, may not be very realistic or adequate. It is necessary to build more complex integrated models that would take into account the likely changes in the socioeconomic conditions of the population and the growth of health infrastructure across the country over the coming decades. In particular the results of anti-malaria measures which are being implemented or have already been planned have to be factored into the models. A major source of uncertainty in future projections about the spread of malaria is how the mosquito vectors may adapt themselves to the global temperature rise. A recent example of an integrated approach is a case study by Garg et al (2009) in which it has been argued that well-crafted and well-managed development policies could help build the resilience of communities and systems and result in reducing the likely adverse impacts of climate change. On the contrary, if the development variables are poorly managed, the impacts of climate change on human health could get aggravated.

It would therefore be prudent for India to design an adaptation strategy that takes cognizance of the uncertainty factors associated with the likely impacts of climate change in the health sector. Such an integrated adaptation strategy would have to include improved surveillance and monitoring systems, preparation of maps for different vectors on small spatial scales, measures for prevention of breeding, medical interventions, and creation of public awareness. All this calls for further research.

A specific feature of the incidence of malaria in India is that locations at altitudes higher than 1.8 km, have been found to be free from it, primarily because of the lower temperatures that prevail there. Such mountain regions need to be a special attention as future climate change may cause the temperature to rise and make them susceptible to malarial outbreaks.

Finally, while the current emphasis on malaria is understandable, the impacts of climate change on other vector-borne and water-borne diseases also need to be given due importance and studies have to be initiated on such other diseases as well.

5.5 References

Aggarwal P. K. and coauthors, 2006, "InfoCrop: A dynamic simulation model for the assessment of crop yields, losses due to pests, and environmental impact of agro-ecosystems in tropical environments. I. Model description", *Agricultural Systems*, 89, 1-25, II. Model performance, *Agricultural Systems*, 89, 47-67.

Bhattacharya S., Sharma C., Dhiman R. C. and Mitra A. P., 2006, "Climate change and malaria in India", *Current Science*, 90, 169-175.

Christensen, J. H. and coauthors, 2007, "Regional Climate Projections", *Climate Change 2007: The Physical Science Basis. Contribution of Working Group I to the Fourth Assessment Report of the Intergovernmental Panel on Climate Change* [Ed: Solomon S. et al], Cambridge University Press,648-940.

Dhar O. N. and Mandal B. N, 1981, "Greatest observed one-day point and areal rainfall of India", *Pure and Applied Geophysics*, 119, 922-933.

Dhiman R. C. 2009, "Climate change and vector-borne diseases in India", *Symposium on Weather, Climate and Sustainable Development,* India Meteorological Department, New Delhi, 17-18 December 2009.

Garg A., Dhiman R. C., Bhattacharya S. and Shukla P. R., 2009, "Development, malaria and adaptation to climate change: A case study from India", *Environmental Management, Springer Online.*

Guhathakurta P. and Rajeevan M., 2006, "Trends in the rainfall pattern over India", *Research Report No. 2/2006,* National Climate Centre, India Meteorological Department, Pune, 25 pp.

Halley E., 1686, "An historical account of the trade winds and monsoons, observable in the seas between and near the tropics with an attempt to assign the physical cause of the said winds", *Philosophical Transactions, Royal Society*, London, 16, 153-168.

IMD 1975, *Hundred Years of Weather Service (1875-1975)*, India Meteorological Department, Pune, 207 pp.

Jain S. L. and coauthors, 2008, "Trend analysis of total column ozone over New Delhi, India", *Mapan - Journal of Metrology Society of India*, 23, 63-69.

Kalra N. and coauthors, 2007, "Impacts of climate change on agriculture", *Outlook on Agriculture*, 36, 109-118.

Kelkar R. R., 2005, "Understanding the extreme weather events", *Newsletter Indian Water Resources Society.*

Kelkar R. R., 2009, *Monsoon Prediction,* BS Publications, Hyderabad, India, 234 pp.

Lemke P. and coauthors, 2007, "Observations: changes in snow, ice and frozen ground", *Climate Change 2007: The Physical Science Basis. Contribution of Working Group I to the Fourth Assessment Report of the Intergovernmental Panel on Climate Change* [Ed: Solomon S. et al], Cambridge University Press, 338-383.

Le Treut H. and coauthors, 2007, "Historical Overview of Climate Change", *Climate Change 2007: The Physical Science Basis. Contribution of Working Group I to the Fourth Assessment Report of the Intergovernmental Panel on Climate Change* [Ed: Solomon S. et al], Cambridge University Press, 94-127.

Mani N. J., Suhas E. and Goswami B. N., 2009, "Can global warming make Indian monsoon weather less predictable?", *Geophysical Research Letters*, 36, L08811, doi:10.1029/2009GL037989.

MoEF, 2009a, "GHG inventory information", *India's Initial National Communication,* Ministry of Environment and Forests, Government of India, Chapter 2, 31-56.

MoEF, 2009b, "Vulnerability assessment and adaptation", *India's Initial National Communication,* Ministry of Environment and Forests, Government of India, Chapter 3, 59-132.

Mukhopadhyay P., Taraphdar S., Goswami B. N. and Krishna Kumar K., 2009, "Indian summer monsoon precipitation climatology in a high resolution regional climate model: Impact of convective parameterization on systematic biases", *Weather and Forecasting*, doi:10.1175/ 2009WAF2222320.1 (IF 1.375)

Ortiz R. and coauthors, 2008, "Climate change: can wheat beat the heat?", *Agriculture , Ecosystems and Environment*, 126, 46-58.

Preethi B., Kripalani R. H. and Krishna Kumar K., 2009, "Indian summer monsoon rainfall variability in global coupled ocean-atmospheric models", *Climate Dynamics*, DOI 10.1007/s00382-009-0657-x, 1-19 (IF 3.961)

Rakhecha P. R. and coauthors, 1990, "Homogeneous zones of heavy rainfall of 1-day duration over India, *Theoretical and Applied Climatology*, 41, 213-219.

Ramakrishna Y. S. and coauthors, 2009, "Role of agrometeorology in the context of climate change and variability", *Mausam*, 60, Diamond Jubilee Volume, 101-114.

Singh P., Arora M. and Goel N. K., 2005, "Effect of climate change on runoff of a glacierized Himalayan basin", *Hydrological Processes*, 20, 1979-1992.

Solomon S. and coauthors (Ed), 2007, *Technical Summary, Climate Change 2007: The Physical Science Basis. Contribution of Working Group I to the Fourth Assessment Report of the Intergovernmental Panel on Climate Change,* Cambridge University Press, 91 pp.

WMO, 2009, "WMO statement on the status of the global climate in 2008", *WMO No. 1039,* World Meteorological Organization, Geneva, 13 pp.

Chapter 6

Politics and Economics of Climate Change

While global warming and climate change are essentially a matter of scientific observation, research and prediction, they touch the lives of every human being and the economies of every nation on earth. It is an observed fact that the concentrations of greenhouse gases in the earth's atmosphere have been increasing at an unprecedented rate since the beginning of the industrial era. It is also accepted that unless future global greenhouse emissions are curbed, global warming will continue unabated and will eventually lead to disastrous consequences for the earth and humankind. The problem that arises here is that the science of climate change is not a very definitive one. There are a host of uncertainties yet to be resolved and the predictions of future climate change are not guaranteed to come true. Notwithstanding the lack of a solid scientific basis, it is imperative for actions to be initiated now for what is called saving the planet. This leads to the question as to whose responsibility it is to save the planet.

It is here that politics and economics enter the scene. It is natural for the poorer countries to fix the blame on the developed countries for having brought about the present state of the environment and to ask them to take actions for rectifying the situation. The developed countries are in no mood to do so on their own and they want all the nations of the world to share the burden. The developing countries on the other hand, do not want to reduce their GHG emissions at the expense of their development efforts, nor are they in a position to bear the costs of reducing carbon emissions.

6.1 Historical Responsibility

It is clear that the historical responsibility for the current problem of global warming and climate change lies with the developed countries. Countrywise data on CO_2 emissions are available since 1850 and the cumulative total emission can be computed to obtain an indication of how much each country has contributed in the past to the creation of the present global warming problem. If the cumulative CO_2 emissions over the period 1850-2000 are ranked countrywise, the U. S. comes first, followed by the European Union, with China ranking fourth and India ninth (Table 6.1.1). Mexico ranks sixteenth and the rest of the countries have less than 1 % share of the global total. There is a sharp fall in the percentage values after the two top emitters, U. S. and European Union. The developing countries considered together, have contributed only 24 % of global cumulative emissions whereas the developed countries have had a share of 76 % in the historical emissions.

Table 6.1.1 Cumulative CO_2 emissions of different countries expressed in terms of percentage of the total global emission
(Source: World Resources Institute, Washington DC, USA
web site http://www.wri.org)

Rank for 1850-2000	Country	Share of total global emission (%)		
		1850-2000	1950-2000	1990-2000
1	United States	29.3	26.4	23.5
2	EU-25	26.5	22.0	17.0
3	Russia	8.1	9.5	7.6
4	China	7.6	9.0	13.8
5	Germany	7.3	-	-
6	United Kingdom	6.3	-	-
7	Japan	4.1	-	5.2
8	France	2.9	-	-
9	India	2.2	-	3.8
10	Ukraine	2.2	-	-
11	Canada	2.1	-	-
12	Poland	2.1	-	-
13	Italy	1.6	-	-
14	South Africa	1.2	-	-
15	Australia	1.1	-	-
16	Mexico	1.0	-	-

Historical estimates like those given in Table 6.1.1 are likely to be influenced by several factors, like the chosen methodology of calculation, unavailability of official data for some countries and lack of even unofficial data for many countries, particularly in respect of non-CO_2 GHG emissions. If only recent data are taken into account, the numbers change and the current share of many countries is at variance with their historical share. So such data cannot serve as the basis of a legal agreement that would be internationally acceptable. However, even if more recent periods are taken for the analysis, say 1950-2000 or 1990-2000, the emissions of the U. S. and the European Union continue to rank the highest and second highest in the world respectively. India's current share remains very low at 3.8 %, which is just a little higher than its historical share of 2.2 %.

The climate change debate which began about 20 years ago has since snowballed and it will not be an exaggeration to say that the science of climate change is the only physical science, apart from nuclear science, to have had such tremendous and far-reaching political and economic ramifications. The following sections describe the various measures adopted by the international community with regard to climate change, the agreements and disagreements, and the results of actions taken so far.

6.2 Montreal Protocol

The first ever international agreement in the field of climate change and the protection of the environment was the so-called Montreal Protocol. In 1985, very soon after the Antarctic ozone hole had been discovered and the underlying physical and chemical processes and anthropogenic causes had been understood, a treaty called the Vienna Convention for the Protection of the Ozone Layer was signed by 20 nations in Vienna. They agreed under this treaty to take appropriate measures to protect the earth's ozone layer from human activities. Subsequently, the Montreal Protocol on Substances that Deplete the Ozone Layer was signed in 1987 and it came into force in 1989.

The Montreal Protocol established legally binding controls on developed as well as developing nations on the production and consumption of halogen source gases known to cause ozone depletion. The 1990 London Amendments to the Montreal Protocol stipulated a phasing out of the production and consumption of the most damaging ozone-depleting substances in developed nations by 2000 and in developing nations by 2010. The 1992 Copenhagen Amendments advanced the deadline for developed countries to 1996 and also brought new substances under regulation. Further controls on ozone-depleting substances were introduced in the amendments agreed upon in 1997 at Montreal and in 1999 at Beijing. India ratified the amendments in 2003.

As a result of the Montreal Protocol, the total abundance of ozone-depleting gases in the atmosphere has begun to decrease in recent years. If the nations of the world continue to follow the provisions of the Montreal Protocol, the decrease will continue throughout the 21st century. Some individual gases, such as halons and hydro-chloro-fluoro-carbons (HCFCs), are still increasing in the atmosphere but will begin to decrease in the next decades if compliance with the Protocol continues. Around the middle of the 21st century, the effective abundance of ozone-depleting gases should fall to values that were present prior to the formation of the Antarctic ozone hole.

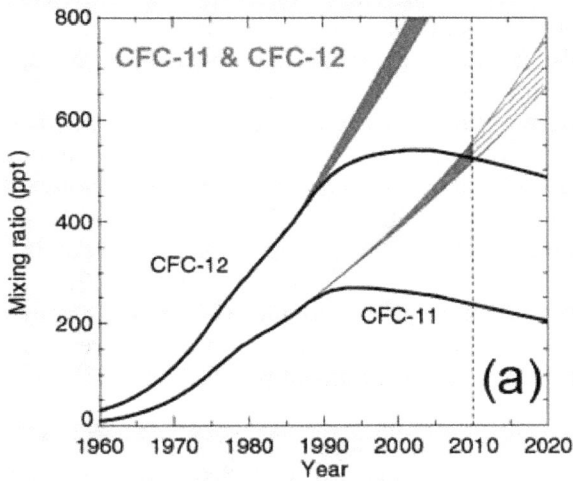

Baseline ODS conditions as measured in the past and projected for the future.

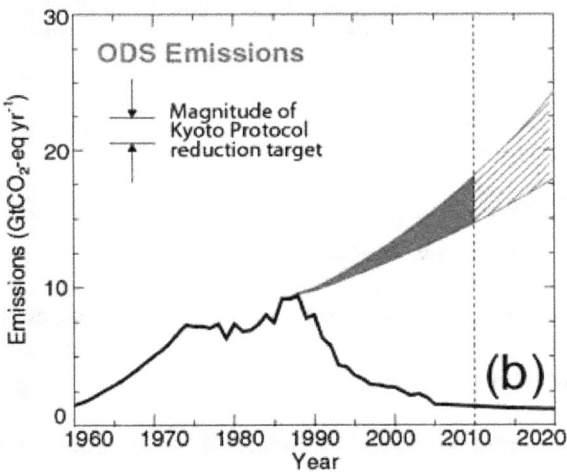

ODS projections for a world with no regulations from the Montreal Protocol.

Figure 6.2.1 Decreasing abundance of (a) ozone-depleting substances in the atmosphere as a result of the Montreal Protocol and (b) projected unregulated increase without it (Source: WMO 2008)

It is important to note that the ozone depleting substances (ODS) are also greenhouse gases by themselves. Therefore, a reduction in the concentrations of ODS in the atmosphere not only helps to restore the depleting ozone in the atmosphere, but also reduces their warming potential. Figure 6.2.1 (a) shows how measured atmospheric abundances of CFC-11 and CFC-12 (black curves) started declining in the early 1990s. They are currently less than a half of what they would have been without the Montreal Protocol (red curves). Figure 6.2.1 (b) shows the sum of all ODS emissions expressed as an equivalent emission of CO_2. Starting in the late 1980s, ODS emissions fell sharply (black curve) compared to those expected without the Montreal Protocol (red curve). It is estimated (WMO 2008) that by 2010, the Montreal Protocol will have resulted in the reduction of ODS emissions by an amount equivalent to 11 Gton CO_2 eq/yr which is 5 to 6 times the reduction target of 2 Gton CO_2 eq/yr in the first commitment period (2008-2012) of the Kyoto Protocol.

Substantial recovery of the ozone layer is expected near the middle of the 21st century, assuming global compliance with the Montreal Protocol. Recovery will occur as chlorine- and bromine-containing gases that cause ozone depletion decrease in the coming decades under the provisions of the Protocol. However, the influence of future changes in climate and other atmospheric parameters or events like volcanic eruptions could modify the rate of ozone recovery over several years,

6.2.1 India and the Montreal Protocol

India acceded to the Vienna Convention for the Protection of the Ozone Layer in 1991. It signed the Montreal Protocol along with its London Amendment in 1992 and also ratified the Copenhagen, Montreal and Beijing Amendments in 2003. India has since remained committed towards fulfilling its responsibility as a Party to the Montreal Protocol and has always been mindful of its responsibility in this global effort.

India's compliance to the Montreal Protocol has been achieved through a combination of regulatory, fiscal and trade-related measures and investments in technology transfer, technical assistance, training and capacity building. The Government of India has offered exemption from payment of customs and central excise duties for import of goods meant for technologies which do not use ozone depleting substances, and licensing systems for import and export of CFCs and halons. The Government has entered into agreements with certain manufacturing enterprises for reduction of CFCs production in a specific phased manner. It is also taking actions against illegal trading in banned substances (MoEF 2003).

All these steps have resulted in India having achieved the target of phasing out CFCs and other ozone depleting substances in August 2008 which is 17 months ahead of the target date of 1 January 2010. However, a new factor that has now assumed importance and has become a cause for concern is that the HCFCs which replaced the CFCs have been found to contribute to global warming. Hence, steps have to be taken to phase them out.

However, this is a particularly difficult task for India, the reason being that due to the recent economic and industrial growth, the consumption of HCFCs in India has increased steadily at an average annual rate of over 11 % in the past 15 years. Much of this increase has occurred in the past few years and since 2001, the consumption of HCFCs in India has more than tripled. This trend is expected to continue. The accelerated phase-out schedule for HCFCs accepted by the Parties to the Montreal Protocol therefore poses unprecedented challenges to India. The challenge lies in eliminating the use of HCFCs by finding alternative technologies that are both ozone- and climate-friendly but without overburdening the Indian economy. India has unveiled concrete measures targeted at freezing the use of HCFCs at 2009 levels by 2013 and gradually bringing it to zero by 2030 (MoEF 2009).

6.3 IPCC

Policymakers find climate change to be a very complex issue and they need an objective source of information about the causes of climate change, its potential environmental and socio-economic consequences and the adaptation and mitigation options to respond to it. To meet this need, the Intergovernmental Panel on Climate Change (IPCC) was established in 1988 jointly by the World Meteorological Organization (WMO) and the United Nations Environment Programme (UNEP). The task of the IPCC is to assess the scientific, technical and socioeconomic information relevant for understanding the risk of human-induced climate change. Its specific mandate is:

(a) Identification of uncertainties and gaps in our present knowledge with regard to climate change and its potential impacts, and preparation of a plan of action over the short-term in filling these gaps;

(b) Identification of information needed to evaluate policy implications of climate change and response strategies;

(c) Review of current and planned national/international policies related to the greenhouse gas issue;

(d) Scientific and environmental assessments of all aspects of the greenhouse gas issue and the transfer of these assessments and other relevant information to governments and intergovernmental organizations to be taken into account in their policies on social and economic development and environmental programs.

The IPCC has three Working Groups, a Task Force and a Task Group. Working Group I (WG I) assesses the physical scientific aspects of the climate system and climate change. These include changes in GHGs and aerosols, air, land and ocean temperatures, rainfall, glaciers and ice sheets, oceans and sea level, climate models, climate projections and causes and attribution of climate change. Working Group II (WG II) assesses the vulnerability of socio-economic and natural systems to climate change, consequences of climate change, and adaptation options. It also takes into consideration specific sectors like water resources, ecosystems, food, forests, coastal systems, industry and human health and also specific regions of the world. Working Group III (WG III) assesses options for mitigating climate change through limiting or preventing greenhouse gas emissions and enhancing activities that remove them from the atmosphere. It considers sectors like energy, transport, buildings, industry, agriculture, forestry and waste management.

The IPCC Task Force on National Greenhouse Gas Inventories (TFI) was established by the IPCC to oversee the IPCC National Greenhouse Gas Inventories Programme. Its core activity is to develop and refine an internationally-agreed methodology and software for the calculation and reporting of national GHG emissions and removals. The Task Group on Data and Scenario Support for Impacts and Climate Analysis (TGICA) was established to facilitate cooperation between the climate modelling and climate impacts assessment communities. It aims at facilitating wide availability of climate change related data and scenarios for climate analysis and research. One of its main activities is the coordination and oversight of the IPCC Data Distribution Centre (DDC).

While the IPCC can be described as a scientific body, it is not an independent body of scientists. It does not conduct any research on its own, nor does it directly sponsor research by others. The IPCC does not itself monitor the earth's climate. The assessment reports that it generates from time to time are based on scientific inputs provided by hundreds of scientists from all over the world who contribute to the work of the IPCC as authors, contributors and reviewers. However, as membership of the IPCC is open to all member countries of WMO and UNEP, governments can participate in plenary Sessions of the IPCC where reports are reviewed and accepted. When governments accept the IPCC reports and approve their Summary for Policymakers, they acknowledge the legitimacy of their scientific content.

The IPCC provides its reports at regular intervals. So far, four IPCC Assessment Reports have been brought out in 1990, 1995, 2001 and 2007 respectively. The First Assessment Report led to the adoption of the United Nations Framework Convention on Climate Change (UNFCCC), which was opened for signature in the Rio de Janeiro Summit in 1992 and entered into force in 1994. The Second Assessment Report provided key inputs for the negotiations of the Kyoto Protocol in 1997.

The First Assessment Report of WG I was completed in 1990. Most of its conclusions were qualitative and scientific uncertainty was hardly mentioned. The Second Assessment Report of 1995 contained intensive chapters on the carbon cycle, atmospheric chemistry, aerosols and radiative forcing. It clearly said that the balance of evidence suggested a discernible human influence on global climate. A range in the mean surface temperature increase since 1900 was given as 0.3 °C to 0.6 °C with no explanation about the likelihood of this range. The Third Assessment Report (TAR) of WG I was released in 2001. It said that there was new and stronger evidence that most of the warming observed over the last 50 years was attributable to human activities.

The Fourth Assessment Report (AR4) of WG I was released in 2007 and it contains voluminous information on the current state-of-art of the physical science basis of climate change.

The IPCC has initiated the process of preparing its Fifth Assessment Report (AR5). The outlines and schedules have already been agreed upon. IPCC is currently looking for experts who can act as authors and review editors for the contributions of the three Working Groups to the AR5. The Working Group I report is scheduled to be finalized in September 2013, the Working Group II report in March 2014 and the Working Group III report in April 2014. The scope and content of the AR5 Synthesis Report will be developed in the course of the year 2010. The Synthesis Report is scheduled to be finalized in September 2014.

The IPCC brings out Special Reports from time to time. The Special Report of 1999 on Aviation and the Global Atmosphere was the first complete assessment of an industrial sub-sector. The summary related aviation's role relative to all human influence on the climate system. The Special Reports on

Emissions Scenarios have been widely used for climate modelling experiments. Two Special Reports are currently under preparation by the IPCC. The Special Report on "Renewable Energy Sources and Climate Change Mitigation" will be released in 2010 and it aims to provide a better understanding and broader information on the mitigation potential of renewable energy sources. This report will cover different sources of renewable energy like bio-energy, direct solar energy, geothermal energy, hydropower, ocean energy and wind energy. It will discuss technological and socio-economic aspects, impacts on energy security, opportunities and synergies, and integration of the renewable sources into the energy supply systems. The second Special report is on "Managing the Risks of Extreme Events and Disasters to Advance Climate Change Adaptation" and is planned to be released in 2011. This will consider different types of extreme events and will focus on managing the risk at different levels of society.

The Nobel Peace Prize for 2007 was awarded jointly to the IPCC and Al Gore for their efforts to build up and disseminate greater knowledge about man-made climate change, and to lay the foundations for the measures that are needed to counteract such change.

The citation for the Nobel Peace Prize cautioned that indications of changes in the earth's future climate must be treated with the utmost seriousness as extensive climate changes may alter and threaten the living conditions of mankind. They may induce large-scale migration and lead to greater competition for the earth's resources. Such changes will place particularly heavy burdens on the world's most vulnerable countries. There may be increased danger of violent conflicts and wars, within and between states.

The citation further said that through the scientific reports it has issued over the past two decades, the IPCC has created an ever-broader informed consensus about the connection between human activities and global warming. Thousands of scientists and officials from over one hundred countries have collaborated to achieve greater certainty with respect to the scale of the warming. Whereas in the 1980s global warming seemed to be merely an interesting hypothesis, the 1990s produced firmer evidence in its support. In the last few years, the connections have become even clearer and the consequences still more apparent.

The citation hoped that the award of the Nobel Peace Prize for 2007 to the IPCC and Al Gore would bring to a sharper focus on the processes and decisions that appear to be necessary to protect the world's future climate, and thereby to reduce the threat to the security of mankind. Action is necessary now, it warned, before climate change moves beyond man's control.

6.3.1 A Critique of IPCC's Work

IPCC has been providing the basic scientific framework and background for the continuing climate change negotiations among the countries of the world leading to economic and political agreements of far-reaching implications, including the Kyoto Protocol. The award of the Nobel Peace Prize is an acknowledgement of the pivotal role played by IPCC in this regard. However, a clear distinction needs to be made that this is a prize for Peace, or for the contribution of the organization towards the furtherance of global well-being, and it is not an endorsement of the science of climate change. Had it been so, the Nobel Prizes for Physics or Chemistry would have been awarded to the scientists working in this field.

In fact many scientists have questioned the IPCC projections of future climates and there is an emerging dissenting view (for example, Khandekar et al 2005) on the causes and consequences of global warming as put forth by the IPCC. Debate is an essential ingredient of the scientific process, and is common to all branches of science, but in the case of climate change, when some exaggerated and unfounded claims are made, it is not only the science but also the methodology of the IPCC assessments, that comes under criticism.

A recent case in point was about the recession of Himalayan glaciers. In 1999, the *New Scientist* magazine had reported a prediction that the Himalayan glaciers were fast retreating and could disappear by the year 2035 (Pearce 1999). This prediction was made over an email interview given to the magazine and it was never published later in a peer-reviewed journal (Pearce 2010). It was however quoted in IPCC AR4 in Section 10.6.2 of the WG II Report and again in Box TS.6 of the WG II Technical Summary. While the IPCC AR4 had been released in 2007, this statement from the IPCC AR4 suddenly came to the forefront in January 2010, caught global media attention and created an intense controversy. Finally, the IPCC had to intervene and issue an official statement expressing regret for the poor application of standard IPCC procedures in this instance, but maintaining that the threat of glacier retreat was real.

In fact, a paper by Anthwal et al (2006) had also contained a similar statement which was refuted on the basis of a thorough analysis of the rate of retreat of Himalayan glaciers by Jain (2008). He had shown that it may take not 35 years but longer than 700 years for the Gangotri glacier to completely melt. However, this balanced view based on meticulously analysed data and published in a reputed Indian scientific journal, had come after the IPCC AR4 had been published in 2007, wherein the earlier unsubstantiated claim was highlighted.

The issue of Himalayan glaciers can be made out to be an exception, but there has also been some general skepticism about the inner workings of the IPCC and the manner in which IPCC's scientific assessments are made. McKitrick et al (2007) of the Fraser Institute, Vancouver, BC, Canada, have argued that IPCC reports could have an element of bias as chapter authors are frequently asked to summarize current controversies in which they themselves are professionally involved. Chapter authors may also tend to project their own work more prominently than contrary views. Some research that contradicts the hypothesis of greenhouse gas-induced warming may possibly go under-represented, while some controversies may get treated in a one-sided way. While the IPCC enlists many expert reviewers, no indication is given as to how far they agreed or disagreed with the material they reviewed.

As pointed out by McKitrick et al, a more compelling problem is that the Summary for Policymakers which is attached to the IPCC Assessment Report, is produced not by the scientific writers and reviewers alone, but is the result of a process of negotiation with representatives of governments. Their selection of material need not and may not reflect the priorities and intentions of the scientific community itself.

It is interesting to note that the Fraser Institute, Vancouver, BC, Canada, has gone a step ahead and produced its own parallel document called "Independent Summary for Policymakers of the IPCC Fourth Assessment Report" (McKitrick et al 2007) which makes the following points:

- The earth's climate is an extremely complex system and we must not understate the difficulties involved in analyzing it. Despite the many data limitations and uncertainties, knowledge of the climate system continues to advance based on improved and expanding data sets and improved understanding of meteorological and oceanographic mechanisms. The climate in most places has undergone minor changes over the past 200 years, and the land-based surface temperature record of the past 100 years exhibits warming trends in many places.

- Measurement problems, including uneven sampling, missing data and local land-use changes, make interpretation of these trends difficult. Other, more stable data sets, such as satellite, radiosonde and ocean temperatures yield smaller warming trends. The actual climate change in many locations has been relatively small and within the range of known natural variability. There is no compelling evidence that dangerous or unprecedented changes are under way.

- The available data over the past century can be interpreted within the framework of a variety of hypotheses about the causes and mechanisms of the measured changes. The hypothesis that greenhouse gas emissions have produced, or are capable of producing, a significant warming of the earth's climate since the start of the industrial era is credible, and merits continued attention. However, the hypothesis cannot be proven by formal theoretical arguments, and the available data allow the hypothesis to be credibly disputed.

- Arguments for the hypothesis rely on computer simulations, which can never be decisive as supporting evidence. The computer models in use are not, by necessity, direct calculations of all basic physics but rely upon empirical approximations for many of the smaller scale processes of the oceans and atmosphere. They are tuned to produce a credible simulation of current global climate statistics, but this does not guarantee reliability in future climate regimes. And there are enough degrees of freedom in tunable models that simulations cannot serve as supporting evidence for any one tuning scheme, such as that associated with a strong effect from greenhouse gases.

- There is no evidence provided by the IPCC in its Fourth Assessment Report that the uncertainty can be formally resolved from first principles, statistical hypothesis testing or modelling exercises. Consequently, there will remain an unavoidable element of uncertainty as to the extent that humans are contributing to future climate change, and indeed whether such change is good or bad.

It would be worthwhile to seriously consider such comments which are scientifically motivated and based upon scientific principles. They serve to make us aware that the IPCC assessments on the basis of which the world decides its future courses of action, may not quite be the last word on climate change as they are often made out to be.

6.4 UNFCCC

The United Nations Framework Convention on Climate Change was adopted in May 1992 in New York and signed at the 1992 Earth Summit in Rio de

Janeiro by more than 150 countries. It took effect in 1994 when lesser scientific evidence was available about climate change than now. The Convention set its ultimate objective as stabilizing greenhouse gas concentrations at a level that would prevent dangerous anthropogenic (human induced) interference with the climate system. It states that such a level should be achieved within a time-frame sufficient to allow ecosystems to adapt naturally to climate change, to ensure that food production is not threatened, and to enable economic development to proceed in a sustainable manner. For achieving this purpose, the Convention requires precise and regularly updated inventories of greenhouse gas emissions from industrialized countries. The year 1990 has been accepted as the base year for tabulating greenhouse gas emissions. Developing countries also are encouraged to carry out inventories.

Countries ratifying the treaty are called Parties to the Convention and they agree to take climate change into account in such matters as agriculture, industry, energy and natural resources. The Parties agree to develop national programmes to slow climate change. A Conference of Parties (COP) to the UNFCCC has been held every year since 1995. COP-13 was held in December 2007 in Bali, Indonesia, COP-14 in Poznan, Poland in December 2008 and COP-15 in Copenhagen, Denmark in December 2009

The Convention recognizes that it is a framework document and it can be amended or augmented over time so that efforts to deal with global warming and climate change can be focused and made more effective. The first addition to the treaty, the Kyoto Protocol to the UNFCCC, was adopted in 1997 at COP-3 held in 1997 in Kyoto, Japan.

The Convention places the heaviest burden for fighting climate change on industrialized nations, since they are the source of most past and current greenhouse gas emissions. These countries are asked to do their most to cut carbon emissions and also to provide funds for supporting such efforts elsewhere. These developed nations, which are called Annex I countries as well as 12 economies in transition were expected by the year 2000 to reduce emissions to 1990 levels. As a group, they succeeded. Industrialized nations agree under the Convention to support climate change activities in developing countries by providing financial support in addition to what they already provide to these countries. They also agree to share technology with less advanced nations.

Since economic development is vital for the world's poorer countries and because such progress is difficult to achieve even without the complications added by climate change, the Convention accepts that the share of greenhouse gas emissions produced by developing nations will grow in the coming years. It nonetheless seeks to help such countries to limit emissions in ways that will not hinder their economic progress. The Convention acknowledges the vulnerability of developing countries to climate change and calls for special efforts to ease the consequences.

6.5 Kyoto Protocol

The Kyoto Protocol is an international agreement linked to the United Nations Framework Convention on Climate Change (UNFCCC). The major feature of the Kyoto Protocol is that it sets binding targets for 37 industrialized countries and the European community for reducing greenhouse gas (GHG) emissions .These amount to an average of 5 % against the levels prevailing in the base year 1990 over the 5-year period 2008-2012.

The major distinction between the Protocol and the Convention is that while the Convention only encouraged the industrialized countries to stabilize their GHG emissions, the Protocol places a legal binding on them to do so. The Kyoto Protocol recognizes that the developed countries are principally responsible for the current high levels of GHG emissions in the atmosphere as a result of more than 150 years of industrial activity. The Protocol therefore places a heavier burden on the developed nations under the principle of common but differentiated responsibilities.

The Kyoto Protocol was adopted in Kyoto, Japan, on 11 December 1997 and entered into force on 16 February 2005. 184 Parties of the Convention have ratified this Protocol. The detailed rules for the implementation of the Protocol were adopted at COP 7 in Marrakesh in 2001, and have come to be known as the Marrakesh Accords.

The Kyoto Protocol comes to an end in 2012 and a successor agreement needs to be worked out.

6.6 CDM and Carbon Trading

The parties to the UNFCCC Convection were expected to meet the emission reduction targets assigned to them through their own national efforts. The Kyoto Protocol, however, offered them additional means of meeting their targets by way of three market-based mechanisms: (1) emissions trading, (2) Clean Development Mechanism or CDM and (3) joint implementation.

Parties with commitments under the Kyoto Protocol (Annex B Parties) have accepted targets for limiting or reducing emissions. These targets are expressed as levels of allowed emissions, or "assigned amounts," over the 2008-2012 commitment period. The allowed emissions are divided into "assigned amount units" (AAUs). It may happen that some of the Annex B Parties have spare or unused AAUs. Article 17 of the Kyoto Protocol allows such countries to sell this excess capacity to countries that are not able to meet their assigned targets. In effect, the Kyoto Protocol gave GHG emissions the status of a commodity, and gave legitimacy to the buying and selling of this new commodity in what came to be known as the carbon market.

In fact carbon trading in not confined to excess AAUs alone, but much more can be traded under the emissions trading scheme. The other units which may be transferred under the scheme, each equal to 1 tonne of CO2, may be in the form of: Removal units (RMUs) on the basis of land use, land-use change and forestry (LULUCF) activities such as reforestation, emission reduction units (ERUs) generated by a joint implementation project and certified emission reductions (CERs) generated from a clean development mechanism (CDM) project activity. Transfers and acquisitions of these units are tracked and recorded through the registry systems under the Kyoto Protocol. An international transaction log ensures secure transfer of emission reduction units between countries.

In order to address the concern that Parties could "oversell" units, and subsequently be unable to meet their own emissions targets, each Party is required to hold a minimum level of ERUs, CERs, AAUs and RMUs in its national registry. This is known as the commitment period reserve. Emissions trading schemes may be established as climate policy instruments at the national level and the regional level. Under such schemes, governments set emissions obligations to be reached by the participating entities. The European Union emissions trading scheme is the largest in operation.

The Clean Development Mechanism (CDM), defined in Article 12 of the Protocol, allows a country with an emission-reduction or emission-limitation commitment under the Kyoto Protocol (Annex B Party) to implement an emission-reduction project in developing countries. Such projects can earn saleable certified emission reduction (CER) credits, each equivalent to one tonne of CO_2, which can be counted towards meeting Kyoto targets. The mechanism is seen by many as a trailblazer. It is the first global, environmental investment and credit scheme of its kind, providing a standardized emissions offset instrument, CERs.

A CDM project activity might involve, for example, a rural electrification project using solar panels or the installation of more energy-efficient boilers. The mechanism stimulates sustainable development and emission reductions, while giving industrialized countries some flexibility in how they meet their emission reduction or limitation targets.

A CDM project must provide emission reductions that are additional to what would otherwise have occurred. The projects must qualify through a rigorous and public registration and issuance process. Approval is given by the Designated National Authorities. Public funding for CDM project activities must not result in the diversion of official development assistance. The mechanism is overseen by the CDM Executive Board, answerable ultimately to the countries that have ratified the Kyoto Protocol.

Operational since the beginning of 2006, the mechanism has already registered more than 1,000 projects and is anticipated to produce CERs amounting to more than 2.7 billion tonnes of CO_2 equivalent in the first commitment period of the Kyoto Protocol, 2008-2012.

The philosophy of the Clean Development Mechanism (CDM) is that reduction of emissions should not come in the way of sustainable development. CDM enables assistance to developing countries for adopting clean energy and industrial processes including transfer of technology. However, carbon trading has underlying moral issues.

In this newly established carbon market, developed countries can continue with their GHG emissions from their own soil, but claim a discount by helping developing countries to reduce their GHG emissions. By selling carbon, the developing countries stand to benefit from the assistance provided by the countries involved in buying carbon. It is thus an apparent win-win situation, but in reality it is more of an escape clause or a license for the developed countries to carry on with their GHG emissions without restraint.

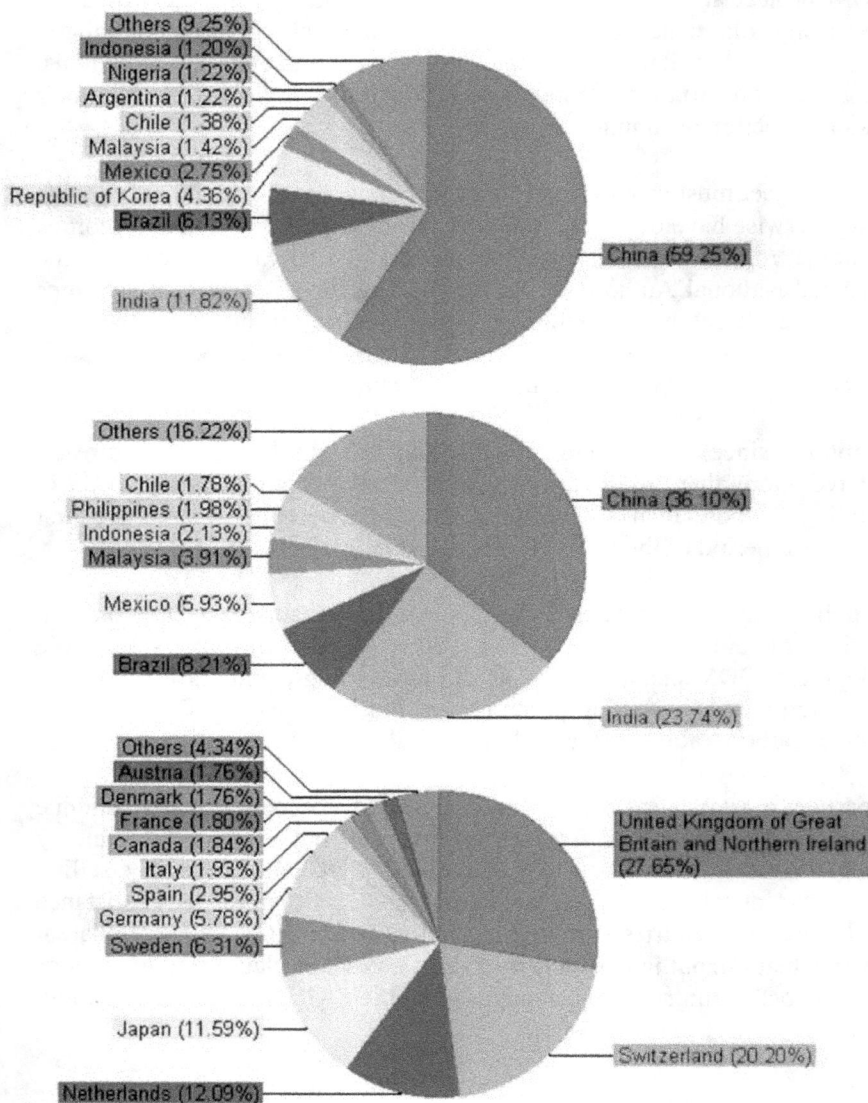

Figure 6.6.1 Statistics of CDM projects: (Top) Registered project activities by host party, (Middle) expected annual average CERs from registered projects of host party, (Bottom) Countrywise breakup of investor parties (Source: CDM web site http://cdm.unfccc.int)

As of January 2010, the registered project activities by host party numbered around 2,000, of which China had 730 or 36 % and India 480 or 24 % (Figure 6.6.1 top). The expected annual average CERs from registered projects of host party were about 340 million of which China had the largest share of 59 % followed by India with about 12 % (Figure 6.6.2 middle). The countrywise breakup of investor parties (Figure 6.6.3 bottom) shows that U. K. has a share of 28 %, Switzerland 20 %, Netherlands and Japan 12 %. The U. S. does not appear in the list of investor countries.

Contrary to earlier fears that carbon trading will be adversely affected by the recent global recession, global market volumes in carbon trading in fact more than doubled in the first half of 2009 compared with the previous year. The European Union Emissions Trading Scheme accounted for the bulk of the growth presumably because of recession-hit industrials selling allowances to raise cash. The EU trading scheme holds as much as three-fourths of the billion dollar global carbon market. Outside the EU, however, the impacts of recession were mixed as the CERs generated under the CDM projects went up but their values fell. There is also a growing uncertainty about the future place of CDM after the Kyoto Protocol comes to an end in 2012.

India is the second top source of carbon offsets under the CDM and traditionally CDM projects were being developed by Indian investors themselves but they are now getting affected by the falling value of CERs and have to find investors. More than a thousand CDM projects in India were formally approved under the CDM mechanism in 2009, one-third of which were biomass projects and another one-third were wind farm projects.

6.7 Outcome of the Kyoto Protocol

As discussed at the beginning of this chapter, the question of equity in emissions reductions is one on which agreement cannot be easily reached. The developed countries are largely held responsible for the current dangerous levels of greenhouse gases in the atmosphere, yet the developing world will likely be hit the hardest by climate change. The demand that developing nations should also be bound by any agreement on emissions reductions is seen as an attempt to stifle their economic and industrial growth

The climate change policy of the United States has been that it will not enter into an agreement to reduce greenhouse gas emissions that is detrimental to its economy or which does not require a meaningful involvement on the part of developing nations. Against this background, the Kyoto Protocol could only call for the reduction of GHG emissions for several industrialized nations to below their 1990 levels by 2008-2012. However, the Protocol also

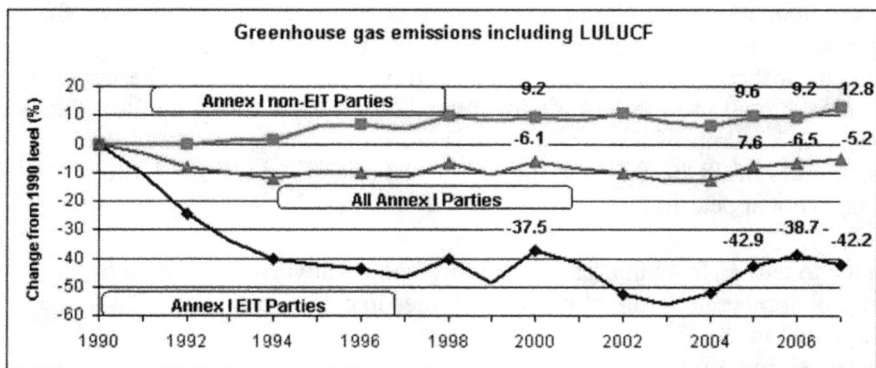

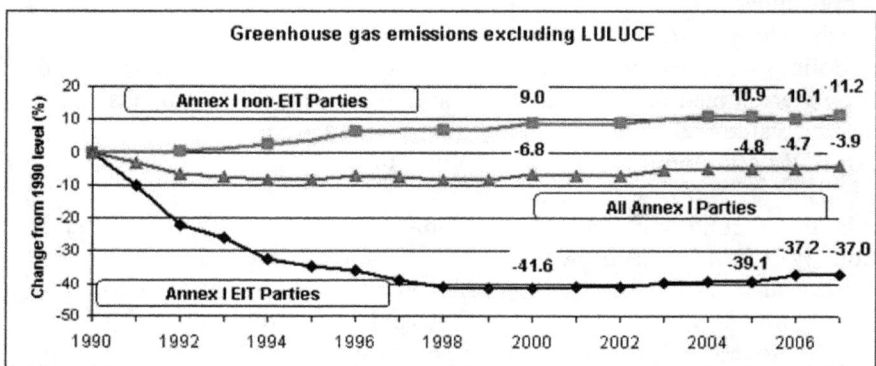

Figure 6.7.1 Trends for 1990-2007 in aggregate GHG emissions of Annex I
Parties. EIT: Economies in Transition, LULUCF: Land Use, Land Use
Change and Forestry (Source: UNFCCC web site http://unfccc.int/ghg_data)

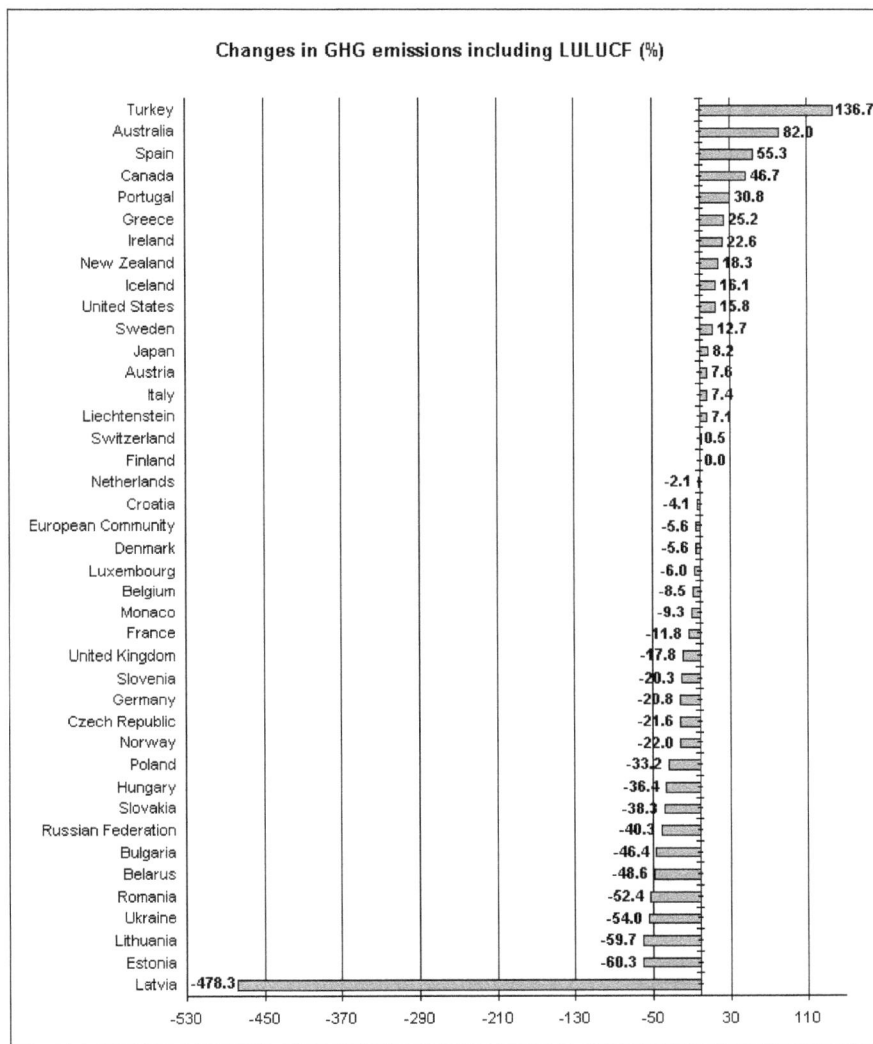

Figure 6.7.2 Changes over 1990-2007 in total aggregate GHG emissions including LULUCF (Land Use, Land Use Change and Forestry) of individual Annex I Parties
(Source: UNFCCC web site http://unfccc.int/ghg_data)

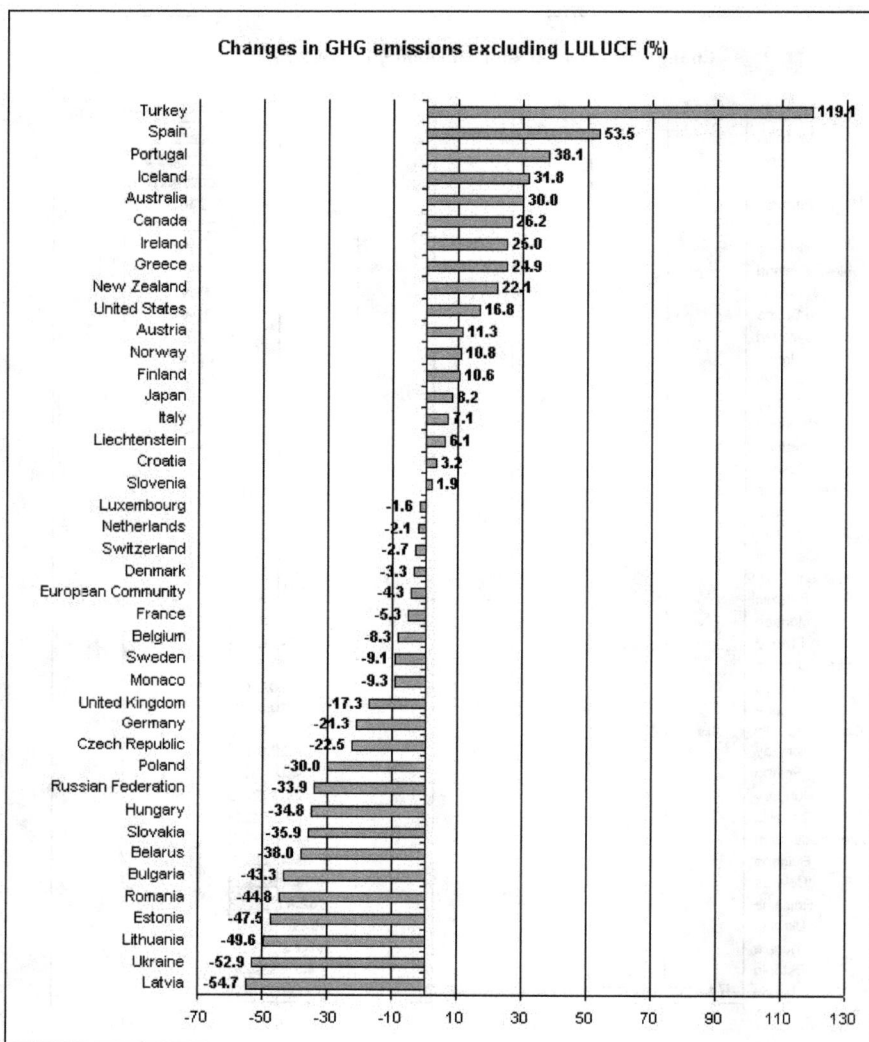

Changes in GHG emissions excluding LULUCF (%)

Turkey	119.1
Spain	53.5
Portugal	38.1
Iceland	31.8
Australia	30.0
Canada	26.2
Ireland	25.0
Greece	24.9
New Zealand	22.1
United States	16.8
Austria	11.3
Norway	10.8
Finland	10.6
Japan	8.2
Italy	7.1
Liechtenstein	6.1
Croatia	3.2
Slovenia	1.9
Luxembourg	-1.6
Netherlands	-2.1
Switzerland	-2.7
Denmark	-3.3
European Community	-4.3
France	-5.3
Belgium	-8.3
Sweden	-9.1
Monaco	-9.3
United Kingdom	-17.3
Germany	-21.3
Czech Republic	-22.5
Poland	-30.0
Russian Federation	-33.9
Hungary	-34.8
Slovakia	-35.9
Belarus	-38.0
Bulgaria	-43.3
Romania	-44.8
Estonia	-47.5
Lithuania	-49.6
Ukraine	-52.9
Latvia	-54.7

Figure 6.7.3 Changes over 1990-2007 in total aggregate GHG emissions
excluding LULUCF (Land Use, Land Use Change and Forestry) of
individual Annex I Parties
(Source: UNFCCC web site http://unfccc.int/ghg_data)

indicated that there would be no binding commitments required of developing countries.

Even if the Kyoto Protocol is fully implemented by all parties it would have resulted in just a 5.2 % reduction of GHG emissions below the 1990 levels. Technically, as per UNFCCC statistics, between 1990 and 2007, the total aggregate GHG emissions including emissions/removals from land use, land use change and forestry (LULUCF) for all Annex I Parties taken together decreased by 5.2 %. If the emissions/removals from LULUCF are excluded, the decrease was 3.9 %. For Annex I Parties with economies in transition (Annex I EIT Parties), GHG emissions including LULUCF decreased by 42.2 % while GHG emissions excluding LULUCF decreased by 37 % (Figure 6.7.1).

The catch here is that for Annex I non-EIT Parties, GHG emissions including LULUCF increased by 12.8 %, and GHG emissions excluding LULUCF increased by 11.2 %. It is also clear from the graphs that wherever there was a decrease in the GHG emissions, it had been accomplished by 1998, after which the emissions had either stabilized or even begun to increase again.

The total change in GHG emissions for individual Annex-I countries over the period 1990-2007 is shown in Figures 6.7.1 and 6.7.2. Excluding LULUCF, the change amounts to an increase of 30 % for Australia, 26.2 % for Canada, 16.8 % for the U. S., 8.2 % for Japan. France, Germany and U. K. have registered a decrease in the GHG emissions.

6.8 India's Greenhouse Gas Inventory

All parties to the UNFCCC are required to prepare and periodically update a national inventory of anthropogenic GHG emissions using comparable methodologies. It is, however, a stupendous task to collect and compile the huge amount of relevant data, especially for countries of the size of India. The latest inventory figures for India's GHG emissions are available for the year 1994 on which India's Initial National Communication to the UNFCCC was based (MoEF 2009a) Table 6.8.1 shows India's aggregate GHG emissions for 1994 for each of the three GHGs separately and Table 6.8.2 gives the sectorwise breakup of India's GHG emissions for the same year.

Table 6.8.1 India's aggregate anthropogenic GHG emissions in 1994
(Source: MoEF 2009a)

GHG	Emission	Unit
CO_2	7,93,490	Gg CO_2-eq
CH_4	18,083	Gg CH_4-eq
N_2O	178	Gg N_2O-eq
Total of all GHGs	12,28,540	Gg CO_2-eq

Table 6.8.2 India's sectorwise GHG emissions in 1994
(Source: MoEF 2009a)

Sector	GHG Emission Gg CO_2-eq	% Share
Energy	7,43,820	61
Agriculture	3,44,485	28
Industrial processes	1,02,710	8
Waste disposal activities	23,233	2
LULUCF	14,292	1
Total of all sectors	12,28,540	100

The GHG emissions in India's energy sector which had a share of 61 % of the total, came mainly from the combustion of fossil fuels. Among the fossil fuels again, the emission from coal combustion was 4,75,530 Gg CO_2-eq or 64 % of the energy sector emissions. The non-CO_2 emissions in this category came from biomass burning, coal mining and handling of oil and natural gas systems. Considering the total emissions by the energy sector, the major contribution of 47 % came from electric power generation, 20 % from industry and 11 % from the transport sector.

Next to energy came the agriculture sector which contributed 28 % of the total CO_2-eq GHG emissions, amounting to 3,44,485 Gg CO_2-eq. However, the GHG emissions in the agriculture sector were mainly in CH_4 and N_2O. Of these 59 % were attributable to enteric fermentation, 23 % to rice cultivation, and the remaining to manure management, fertilizer application to soils and burning of agricultural crop residue. GHG emissions from land use, land-use change and forestry (LULUCF) sector amounted to only 1 % and those from waste disposal activities to 2 % of the total from all sectors.

6.9 India and the World

Figures 6.9.1 and 6.9.2 give comparisons of the per capita CO_2 emission and energy consumption by six countries of the world: U. S., Russian Federation, Japan, Germany, China and India. The data are for three different years 1971, 1990 and 2005/2006.

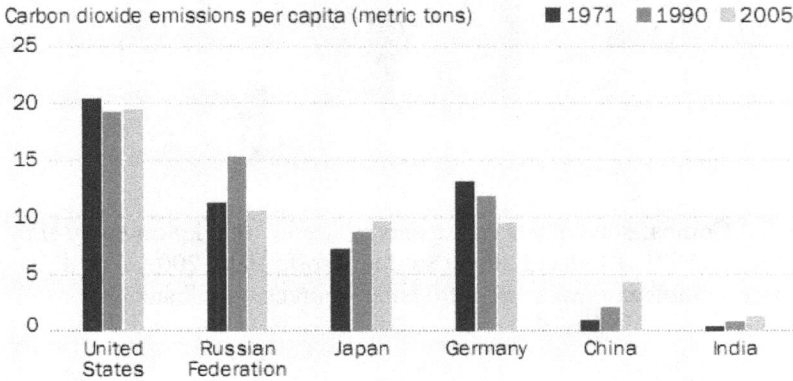

Figure 6.9.1 Comparisons of per capita CO_2 emissions in metric tons by six countries in 1971, 1990 and 2005 (Source: World Bank 2009 World Economic Indicators web site http://siteresources.worldbank.org)

It is said that the top six energy consumers use 55 % of global energy and these six countries include India. However, the figures clearly show that for both per capita CO_2 emission and per capita energy use, the U. S. ranks first and India ranks last and this holds good for all the three years. Moreover, there is a huge difference of 20:1 between the figures for the U. S. and India for CO_2 per capita emissions and 15:1 for per capita energy consumption.

The extreme inequalities that exist among the developed and developing countries have their roots in the history of industrial development and it is sometimes said that the developed countries owe a carbon debt to the rest of the world. The convergence of the carbon footprints across the globe therefore remains a distant dream today and is likely to remain so for many decades to come.

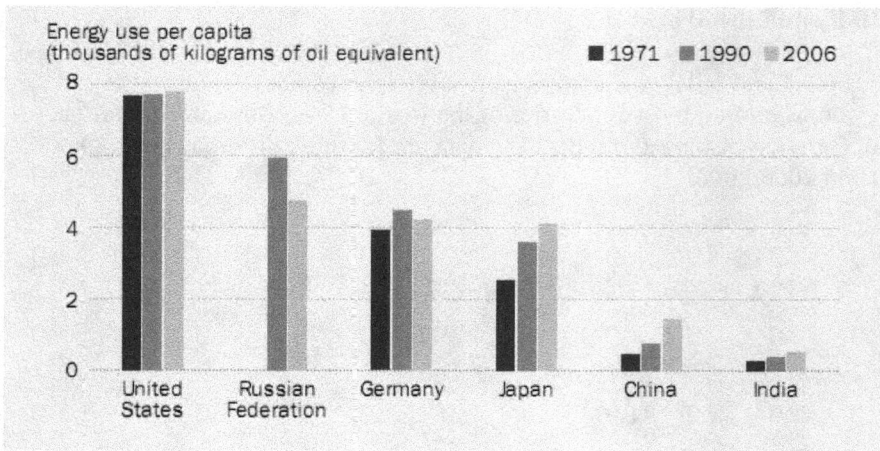

Figure 6.9.2 Comparisons of per capita energy use in 1000 kg oil eq by six countries in 1971, 1990 and 2006 (Source: World Bank 2009 World Economic Indicators web site http://siteresources.worldbank.org)

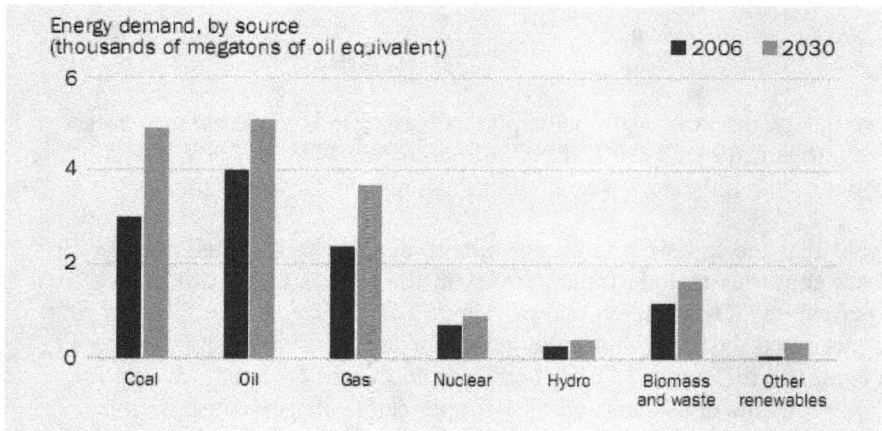

Figure 6.9.3 Comparison of demand for different energy sources in 1000 megatons of oil eq for 2006 and 2030 (Source: World Bank 2009 World Economic Indicators web site http://siteresources.worldbank.org)

The nations of the world require an ever-increasing amount of energy in order to sustain their economic growth, improve the living standards of their rising populations and alleviate poverty. However, there is still a continued and strong dependence on fossil fuels for meeting these energy requirements in spite of the fact that it is causing these resources to deplete rapidly and it is contributing to global warming. Currently, about two-thirds of all global

GHG emissions are attributable to carbon dioxide released through processes related to energy production. What is significant, moreover, is the World Bank projection that even by the year 2030, coal, oil and gas would together be adding up to 80% of the world's energy use, which is the same as today (Figure 6.9.3).

The several projections of the future GHG emissions of different countries that are available are mostly the results of studies carried out by agencies in the developed countries and do not often reflect the ground realities in the developing world. It is therefore significant that the Climate Modelling Forum, India, comprising several independent research institutions in India, has made an indigenous assessment of the scenario based upon five independent modelling studies (CMF 2009).

The five studies use different model structures, methodologies, assumptions, data and definitions of the illustrative scenario. However, four out of the five modelling studies have projected that even if no new GHG mitigation policies are put in place, India's per capita CO_2 emissions in 2030-31 would be between 2.77 and 3.9 tons per capita, which is well below the 2005 global average of 4.22 tons per capita. The fifth study projects the 2031 emissions at 5 tons per capita, which is a little above the 2005 global average. In other words, India's estimated per capita emissions in 2030 are expected to be well below those of the developed countries, even if the developed countries accept 25-40% emission reduction targets.

Further, all the five studies project both the energy intensity and the CO_2 intensity of the Indian economy to fall continuously till 2030-31 in the illustrative scenarios. The CMF (2009) results are reassuring in the context of fears that India's GHG emissions are poised for a runaway increase over the next two decades. The structure of the Indian economy and various policies and measures introduced by the government should keep India on the path of sustainable development without comprising on its responsibilities towards mitigating global climate change

6.10 References

Anthwal A., Joshi V., Sharma A. and Anthwal S., 2006, "Retreat of Himalayan Glaciers – Indicator of Climate Change", *Nature and Science*, 4, 53-59.

CMF, 2009, *India's GHG Emissions Profile Results of Five Climate Modelling Studies,* Climate Modelling Forum, India, Ministry of Environment and Forests, Government of India, 56 pp.

Jain S. K., 2008, "Impact of retreat of Gangotri glacier on the flow of Ganga River", *Current Science*, 95, 1012-1014.

Khandekar M. L., Murty T. S. and Chittibabu P., 2005, "The global warming debate: a review of the state of science", *Pure Applied Geophysics*, 162, 1-30.

MoEF, 2003, *India's Commitment to the Montreal Protocol*, Ministry of Environment and Forests, Government of India, http://envfor.nic.in/news, 4 pp.

MoEF, 2009, *Roadmap for Phase-out of HCFCs in India*, Ministry of Environment and Forests, Government of India, http://moef.nic.in , 116 pp.

MoEF, 2009a, "GHG inventory information", *India's Initial National Communication, Chapter 2*, Ministry of Environment and Forests, Government of India, 31-56.

Pearce F., 1999, "Flooded out", *New Scientist*, 5 June 1999.

Pearce F., 2010, "Debate heats up over IPCC melting glaciers claim", *New Scientist*, 11 January 2010.

WMO, 2008, *WMO Greenhouse Bulletin*, No. 4, World Meteorological Organization, Geneva, 4 pp.

Chapter 7

Preparing for the Future

The United Nations Framework Convention on Climate Change (UNFCCC) now has a near-universal membership of 194 parties. The Kyoto Protocol of the UNFCCC has since been ratified by 190 of the UNFCCC Parties. Under the Kyoto Protocol, 37 states including both highly industrialized countries and economies in transition, have been set with legally binding limitations on their GHG emissions and commitments for their reductions. The UNFCCC and the Kyoto Protocol have a common objective which is to stabilize global greenhouse gas concentrations in the atmosphere at such a level as to prevent their dangerous human interference with the climate system.

The Conference of Parties to the UNFCCC which ended at Bali on 15 December 2007 adopted an action plan which has come to be known as the Bali Roadmap. This consists of a number of forward-looking decisions, the launching of a new negotiation process and the launching of the Adaptation Fund. The Bali Roadmap speaks of climate change as the defining human development challenge of the 21st century.

The negotiation process culminated at the Copenhagen Conference in December 2010.

7.1 Copenhagen Accord

The United Nations Climate Change Conference at Copenhagen was attended by as many as 119 world leaders. It is said to have been the largest gathering of heads of state and government in the history of the U. N. It ended on 19 December 2009 with an agreement by all countries which has come to be known as the Copenhagen Accord.

The Copenhagen Accord is neither a legal agreement nor a treaty nor a protocol. It is essentially a political agreement. However, its significance lies in that it is an expression of the consensus of the political will to combat the effects of climate change. The highlights of the Copenhagen Accord are given below:

- The Accord recognizes the scientific view that global warming should be restricted to less than 2 °C in order to avoid its disastrous consequences.

- It agrees that there should be cooperation towards achieving the peaking of global and national emissions as soon as possible but that this should be achieved on the basis of equity and in the context of sustainable development. It also recognizes that the time frame for peaking will be longer in developing countries.

- It accepts that social and economic development and poverty eradication are the first and overriding priorities of developing countries and that a low-emission development strategy is indispensable to sustainable development.

- The Accord recognizes the critical impacts of climate change and the potential impacts of response measures on countries particularly vulnerable to its adverse effects, and it stresses the need to establish a comprehensive adaptation programme.

- It agrees that developed countries shall provide adequate, predictable and sustainable financial resources, technology and capacity building to support the implementation of adaptation action in developing countries.

- Under the Accord, Annex I Parties to UNFCCC commit to implement, individually or jointly, quantified emissions targets for 2020, and submit them to UNFCCC by 31 January 2010.

- Non-Annex I Parties will implement mitigation actions in the context of sustainable development, and submit them UNFCCC by 31 January 2010. Mitigation actions taken by Non-Annex I Parties will be subject to their domestic measurement, reporting and verification the result of which will be reported through their national communications. There will be provisions for international consultations and analysis under clearly defined guidelines that will ensure that national sovereignty is respected. Least developed countries and small island states may undertake voluntary actions.

- Nationally Appropriate Mitigation Actions (NAMAs) seeking international support will be subject to international measurement, reporting and verification.

- Developing countries, especially those with low-emitting economies should be provided incentives to continue to develop on a low-emission pathway.

- The collective commitment by developed countries is to provide new and additional resources approaching US $ 30 billion for the period 2010-2012 with balanced allocation between adaptation and mitigation. Funding for adaptation will be prioritized for the most vulnerable developing countries. The developed countries commit to a goal of mobilizing jointly US $ 100 billion a year by 2020 to address the needs of developing countries. This funding will come from a wide variety of sources, public and private, bilateral and multilateral, including alternative sources of finance.

- A significant portion of such funding would flow through the Copenhagen Green Climate Fund to be established under the Accord.

- A new Technology Mechanism will be established to accelerate technology development and transfer in support of action on adaptation and mitigation that will be guided by a country-driven approach and be based on national circumstances and priorities.

- The Accord calls for an assessment of its implementation to be completed by 2015. This would include consideration of strengthening actions towards attaining the long-term goal of containing the temperature rise to a lower level of 1.5 °C.

The next annual U. N. Climate Change Conference will take place towards the end of 2010 in Mexico City, preceded by a major two-week negotiating session in Bonn, Germany, in mid-2010.

The Copenhagen Accord of 18 December 2009 underlines that climate change is one of the greatest challenges of our time. It emphasizes that there is a strong political will to urgently combat climate change in accordance with the principle of common but differentiated responsibilities and respective capabilities.

7.2 India's National Initiatives

India has always remained fully committed to fulfilling its international obligations towards the mitigation of climate change. However, it has also maintained the position that it will prefer to take climate change mitigation actions in such a manner that they do not become a hindrance in its path of sustainable development. In 2008, India embarked upon a National Action Plan on Climate Change that clearly sets forth its own national priorities in this direction (MoEF 2009a).

Even in the past, India has taken significant and concrete measures from time to time that have contributed to the mitigation of climate change. One example is the Energy Conservation Act of 2001 which was enacted by the government to ensure the efficient use and conservation of energy in the country. The Bureau of Energy Efficiency was established for this purpose.

Another important legislative measure was the enactment of the Electricity Act of 2003 to promote competition in the electricity sector by separating the generation, transmission, distribution and supply of electricity into independent entities. The objective was to help the development of a power system in the country based on optimal utilization of natural resources. The Electricity Act of 2003 requires State Electricity Regulatory Commissions to specify a percentage of electricity that the electricity distribution companies must procure from renewable sources. In addition, the Electricity Regulatory Commissions are also linking tariffs to efficiency enhancement, thus providing an incentive for renovation and modernization.

Coal is the mainstay of India's energy economy, and coal-based power plants account for about two-thirds of the country's total electric generation installed capacity. New plants are being encouraged to adopt more efficient and clean coal technologies. Several reforms have been introduced in the coal sector to place responsibility on both public and private sectors for scientific mining, and ensuring conservation, safety and protection of the environment and to reorient the overall strategy to take into consideration the role of coal in energy security.

The Auto Fuel Policy of the government announced in 2003 covered the issues related to vehicular emissions, cleaner technologies and security of fuel supply. These measures would result in the efficient combustion of fossil fuels in the road transport sector resulting in reduced GHG emissions. In many Indian cities, public vehicles which previously ran on diesel have switched over in a big way to CNG, thus reducing CO_2 emissions.

India's comprehensive National Plan on Climate Change is based on the premise that while India will deal with the global threat of climate change as a responsible and enlightened member of the international community, it will continue to give priority to the economic and social development of its people and the eradication of poverty. India's development pathway aims at using its unique resource endowments in an ecologically sustainable manner.

This is a highly pragmatic and balanced approach that India has adopted. It contrasts with many other global initiatives that tend to concentrate on the threat of climate change without reference to other pressing problems like population growth, illiteracy, poverty and disease that humanity is also facing.

Under the National Action Plan for Climate Change, eight ambitious Climate Missions have been envisaged (Table 7.2.1). Of these eight missions, two are focused on mitigation, five on adaptation and one on building a knowledge platform and infrastructure. In addition to the eight basic climate missions, India has planned another 24 critical initiatives, as they are called, the detailed plans and institutional framework for which are being worked out (Table 7.2.2).

The National Mission on Strategic Knowledge for Climate Change (NMSKCC) is aimed at building a knowledge platform and infrastructure, and creating what could be called a superhighway for exchange and sharing of information and data required for the climate agenda. It envisages the use of collaborative synergies and activities, and making viable investment in all existing and new knowledge capacities. It proposes the establishment of centres of excellence for research in various aspects of climate change and the creation of national data bases for the climate related aspects of ocean, cryosphere, meteorology, land surface, hydrology, agriculture, forestry, socioeconomics and health. This is a huge task and involves the networking of several government departments and institutions.

Institutional mechanisms have been worked out for effective implementation of the National Action Plan, among which is the creation of an Advisory Council on Climate Change, chaired by the Prime Minister himself and with representation from government, industry and civil society who are the key stakeholders. The Prime Minister's Council will provide guidance towards the implementation of the National Plan and also on India's stand in international negotiations on climate change.

Under the Indian Network for Climate Change Assessment (INCCA), a network of 120 research institutions involving 250 scientists has been launched. Repetition of the word "Programme" in this sentence has to be avoided. Hence the change. The Himalayan Glaciers Monitoring Programme is comprehensive enough to scientifically monitor the Himalayan glaciers. India has plans to launch satellites for monitoring greenhouse gases and aerosols and studying the radiation budget of the atmosphere.

Following the Union Government, individual states are now making their own action plans on climate change, Delhi being the first state to do so.

Table 7.2.1 India's Climate Missions
(Source: MoEF 2009b)

No.	Mission	Objective	Responsible Ministry of the Government of India
1	National Solar Mission	20,000 MW of solar power by 2020	Ministry of New and Renewable Energy
2	National Mission for Enhanced Energy Efficiency (EE)	10,000 MW of EE savings by 2020	Ministry of Power
3	National Mission for Sustainable Habitat	EE in residential and commercial buildings, public transport, Solid waste management	Ministry of Urban Development
4	National Water Mission	Water conservation, river basin management	Ministry of Water Resources
5	National Mission for Sustaining the Himalayan Ecosystem	Conservation and adaptation practices, glacial monitoring	Ministry of Science and Technology
6	National Mission for a Green India	6 million hectares of afforestation over degraded forest lands by the end of 12th Plan	Ministry of Environment and Forests
7	National Mission for Sustainable Agriculture	Drought proofing, risk management, agricultural research	Ministry of Agriculture
8	National Mission on Strategic Knowledge for Climate Change	Vulnerability assessment, research and observation, data management	Ministry of Science and Technology

Table 7.2.2 India's Critical Initiatives (Source: MoEF 2009b)

Type	Initiative
Energy Efficiency in Power Generation	Super critical technologies
	Integrated Gasification Combined Cycle (IGCC) Technology
	Natural Gas based Power Plants
	Closed Cycle Three Stage Nuclear Power Programme
	Efficient Transmission and Distribution
	Hydropower
Other Renewable Energy Technologies Programmes	RETs for power generation
	Biomass based popup generation technologies
	Small scale Hydropower
	Wind Energy
	Grid connected systems
	RETs for transportation and industrial fuels
Disaster Management Response to Extreme Climate Events	Reducing risk to infrastructure through better design
	Strengthening communication networks and disaster management
Protection of Coastal Areas	Undertake measures for coastal protection and setting up Early Warning System
	Development of a regional ocean modelling system
	High resolution coupled ocean-atmosphere variability studies in tropical oceans
	Development of a high-resolution storm surge model for coastal regions
	Development of salinity-tolerant crop cultivars
	Community awareness on coastal disasters and necessary action;
Health Sector	Provision of enhanced public health care services and assessment of increased burden of disease due to climate change
Creating appropriate capacity at different levels of Government	Building capacity in the Central, State and other Agencies/Bodies at the local level to assimilate and facilitate the implementation of the activities of the National Plan

7.3 Harnessing Solar Energy

By being careful with our consumption of fossil fuels, we may succeed in making them last longer, while if we are reckless they may faster. However, it is certain that the earth's reserve of fossil fuels will some day be totally exhausted; only that day could come sooner or later depending upon what we do now. Estimates of the available stock of fossil fuels vary and forecasts of their depletion depend upon the consumption rates assumed in the computation. The world's oil reserves, half of which are in the Middle East, are dwindling fast but they are generally projected to last until around 2070. If no new sources of natural gas are found, the current volume of natural gas is likely to last for another 50 years or so. The largest fossil fuel reserves are that of coal and they are spread all over the earth. However, coal mining is a labour-intensive process as coal lies more than 100 m below the earth's surface. Coal reserves are generally expected to last for at least 500 years and possibly up to a 1000 years from now.

India with its abundance of solar and wind energy, has an edge over the developed countries which are situated in the extra-tropical regions of the earth. Considering that our nearest star, the sun, is going to give its light and heat and drive the earth's atmospheric circulation for billions of years to come, it is clear that India's future lies in the exploitation of these two natural resources that are freely available. In fact, these resources are neither new nor non-conventional as they are sometimes described.

India has launched a National Solar Mission, which has now been named as the Jawaharlal Nehru National Solar Mission. This is one of eight different missions envisaged under the National Action Plan on Climate Change and it aims at giving solar energy the place it deserves in the country's future energy scenario. Solar energy has several distinct advantages over other energy sources, the topmost being its abundance, as India is situated in the tropical belt. Although the country's climate is dominated by the monsoon, most places in India get good sunshine for 250-300 days in a year. The total energy received by the country from the sun in a year is estimated to be a staggering 5×10^{15} kWh.

IMD has been maintaining a countrywide network of stations for monitoring solar radiation since 1957 and therefore a long data series is available. Currently there are over 40 stations in this network where the incident solar radiant energy, direct and diffuse, is measured with automatic instruments (Figure 7.3.1). In addition there are more than a hundred stations which have simple sunshine recorders, which measure the hours of bright sunshine and from which the incoming solar radiation can be statistically estimated.

Mani (1980) and Mani and Rangarajan (1982) had compiled the solar radiation data available until then in order to make an assessment of the availability of solar energy over India. IMD has just brought out an updated publication (IMD 2009) that gives detailed maps and data on solar radiant energy over India.

India as a whole receives as much as 7000 MJm^{-2} of global solar radiation in a year or about 19 MJm^{-2} per day on an average but there is some spatial variation. The daily average solar radiant exposure is close to the country average over most parts of western and southern India, but it is higher over the Rann of Kutch and lower over the Kashmir valley and northeast India (Figure 7.3.2).

The seasonal variation in the global solar radiation over India is however much more pronounced. In the summer months of April and May, when the skies are generally cloudfree except for occasional thunderstorms, large parts of the country receive as much as 22 to 26 per day MJm^{-2} (Figure 7.3.3) whereas the monsoon clouds bring down this value to as low as 14 to 16 MJm^{-2} per day in the months of June to September (Figure 7.3.4). In the winter months of January and February, when the sun is over the southern hemisphere, the daily average falls even further down to 12 MJm^{-2} over north India (Figure 7.3.5).

The other main advantage is that solar energy can be tapped in situ or locally, unlike hydropower or nuclear energy which is produced centrally and has then to be distributed over a grid. The exploitation of solar energy is a very clean process that does not pollute the atmosphere as is the case with thermal power generation in which massive amounts of coal is required to be burnt and the smoke let out into the atmosphere. Both coal and oil have to be transported over long distances, while such transportation costs get eliminated in the case of solar energy. Solar energy is again a safe and clean source when compared to nuclear power which also has great potential but has associated requirements of dealing with radiation safety and disposal of radioactive waste material.

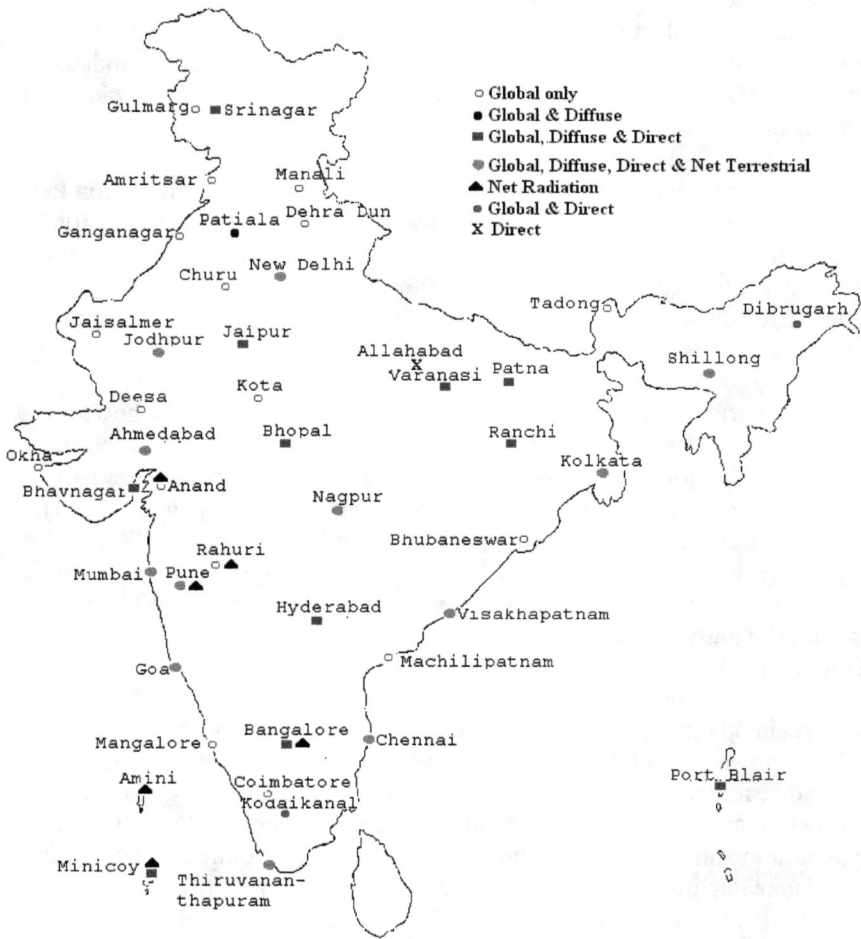

Figure 7.3.1 IMD network of solar radiation monitoring stations
(Source: IMD 2009)

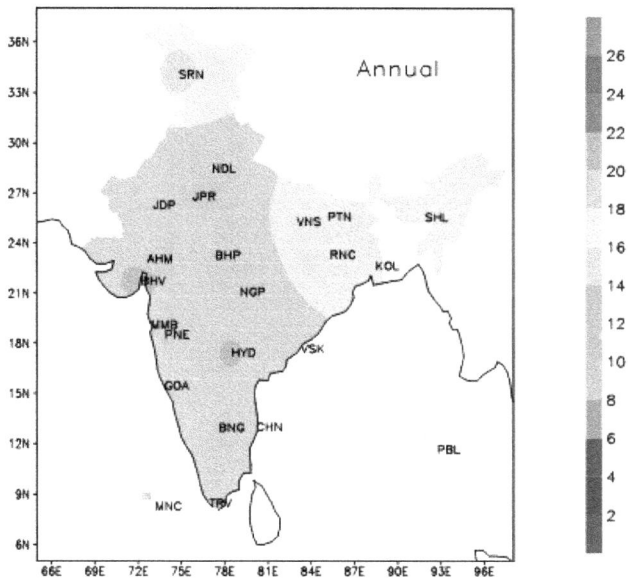

Figure 7.3.2 Annual global solar radiant exposure in MJm^{-2}
(Source: IMD 2009)

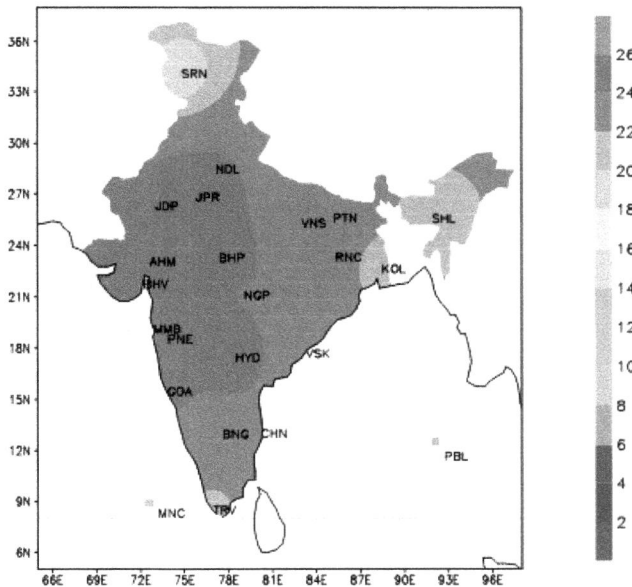

Figure 7.3.3 Global solar radiant exposure in MJm^{-2} for April
(Source: IMD 2009)

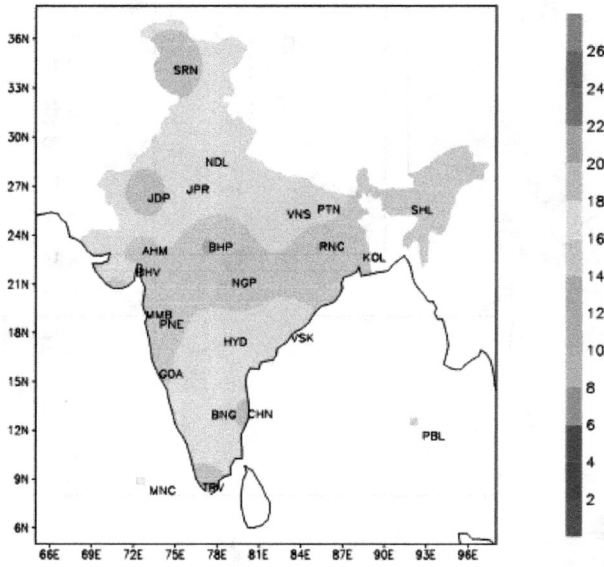

Figure 7.3.4 Global solar radiant exposure in MJm^{-2} for August
(Source: IMD 2009)

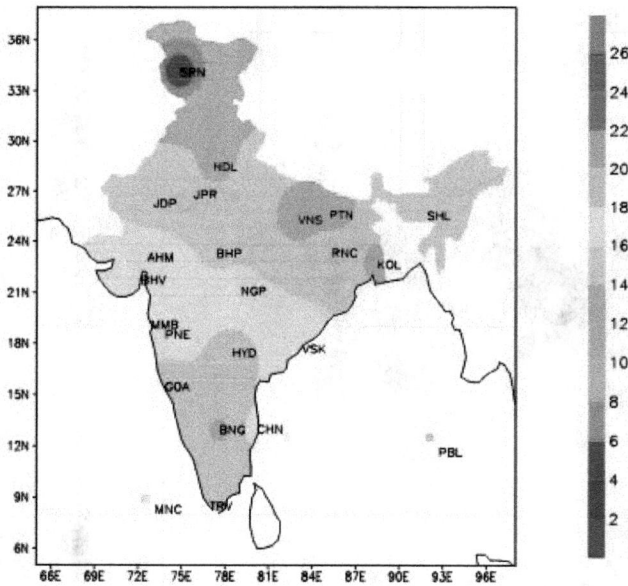

Figure 7.3.5 Global solar radiant exposure in MJm^{-2} for January
(Source: IMD 2009)

Presently the main impediment in the widespread use of solar energy in India is it relatively high production cost. Solar energy can be converted to electrical energy by a process similar to that employed in a thermal power plant or directly by photovoltaic cells. In the thermal type systems, several new methodologies and technologies are being tried and introduced in order to improve the collection and conversion efficiency, for example, parabolic dishes to concentrate the sun's rays, central towers and solar chimneys which drive air draft turbines and do not generate steam. In photovoltaic generation, the semiconductor that is currently in use is silicon dioxide, while bur cadmium telluride is also being tried. The solar cell efficiency needs to be improved considerably in order to make the power generation process commercially viable and affordable to the people. Efforts are already being made to improve the efficiency for example by putting up solar roofing and to make light weight solar cells that can be used in portable devices like solar lanterns. However, the thrust in research and development in this direction needs to be intensified.

The other issue in solar energy exploitation that needs to be specially addressed is its non-availability at night and on many days of the monsoon season, and its reduced intensity in winter. So, paradoxically, while solar energy is generally abundant, it is insufficient or unavailable when it is most required for lighting and heating. For maximum efficiency, the collection systems should have the ability to follow the sun during the day and reorient themselves to the inclination of the sun's beam that changes from day to day. Such dynamic mechanism, however, add considerably to the cost. Therefore, solar energy exploitation must involve the crucial component of storage, enabling the energy to be generated during the most optimum times and then retrieved later as and when required. This is another area in which innovative technologies need to be developed. One way of achieving it at lesser cost is through the adoption of a mix of energies like solar, wind and biomass, so that when one source is absent another may work well. For example during the monsoon season, overcast skies may hide the sun, but additional wind energy could instead be derived from the stronger monsoon winds.

India's National Solar Mission aims at setting up a system of 20,000 MW grid solar power and 2,000 MW of off-grid solar power, including 20 million solar lights and 20 million square metres of solar thermal collectors by 2022. The objective is to create conditions through rapid scale-up of capacity and technological innovation to drive down costs, to achieve grid parity by 2022 and parity with coal-based power by 2030. The Solar Mission will operate in three phases. 2010-2013, 2013-2017 and 2017-2022 for each of which definite targets have been assigned for the production from large size grid interactive power plants, small size and roof top grid interactive power units and decentralized systems, and the area covered by solar thermal collectors

Off-grid photovoltaic solar power generation is well-suited for our islands like Lakshadweep or in remote mountains areas like in Kashmir or Ladakh where there is plenty of sunshine and it is difficult to provide access to an electric power grid. Here, solar power can be locally tapped and used for heating, cooking, laundry, food processing and air-conditioning applications.

7.4 Exploiting Wind Energy

In olden days, sailboats ferried people along rivers while large shipping vessels sailed across oceans with the help of wind power alone. Windmills have been in use for centuries in many countries like Holland, for pumping water or for doing similar manual tasks. However, when we now talk about the exploitation of wind energy, it basically means the conversion of the kinetic energy of the wind into electrical energy. The wind machine that can do this, in its simplest design, consists of a mast, a rotor, a gearbox and a generator. The power of the wind causes the blades of the rotor to turn around the shaft to which they are attached. The shaft is commonly horizontal and the plane of the rotor blades perpendicular to the ground, although other designs with vertical shafts are also used. The shaft transmits power through a series of gears to an electric generator. A wind farm consists of many such units installed over a large area, which can together generate a sizable amount of electric power.

The amount of energy produced by a wind machine depends upon the wind speed and the size of the blades in the machine. As a general rule, as the wind speed doubles, the power increases eight times. Likewise, if the diameter of the circle formed by the blades is doubled, the power increases four times. The individual units that constitute a wind farm have therefore to optimally designed and the location of the wind farm has to be carefully selected in order to derive the maximum benefit from the prevailing winds.

The major limiting factor in wind energy exploitation is the availability of strong, dependable and persistent winds. India is the land of the monsoons, which are essentially an annual reversal of the wind patterns, and the rains that they bring are just a byproduct of the process. During the monsoon season, the winds are strong but when the monsoon withdraws, the wind direction changes significantly and the wind speed drops down. The Centre for Wind Energy Technology (C-WET), Chennai, an autonomous institution under the Ministry of New and Renewable Energy (MNRE) of the Government of India, has prepared a map of India that shows the regions of the country which are rich in wind energy resource (Figure 7.4.1). It can be seen that large parts of the country are not well-suited for the exploitation of

However, Saurashtra and Kutch, west Rajasthan and most of the southern peninsula are potentially ideal zones for this purpose.

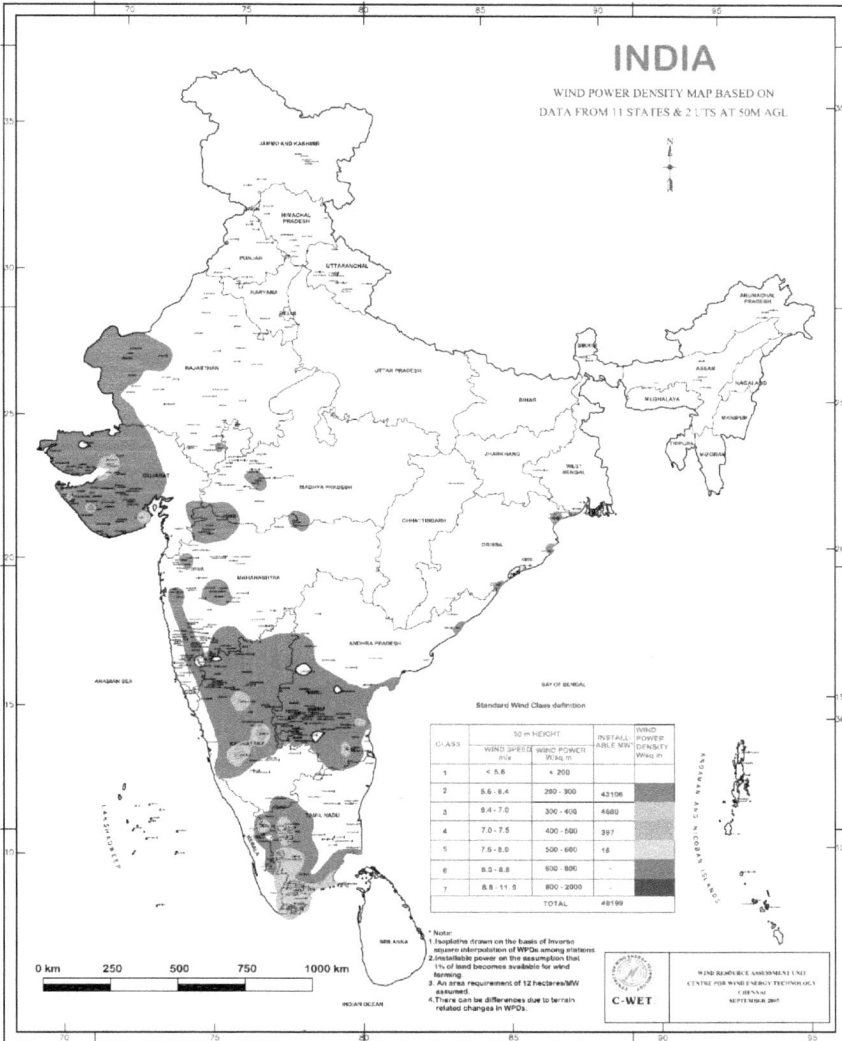

Figure 7.4.1 Wind power density map of India as of September 2005
(Source: C-WET web site http://www.cwet.tn.nic.in)

Table 7.4.1 Estimated Wind Power Potential in various States of India and Installed Capacity (Source: Wind Power India web site http://www.windpowerindia.com)

State	Estimated Wind Power Potential (MW)	Installed Capacity as on 2008 end (MW)
Andhra Pradesh	8275	122.45
Gujarat	9675	1432.71
Karnataka	6620	1184.45
Kerala	875	23.00
Madhya Pradesh	5500	187.69
Maharashtra	3650	1837.85
Orissa	1700	-
Rajasthan	5400	670.97
Tamil Nadu	3050	4132.72
West Bengal	450	1.10
Others	-	3.2
Total for Country	45195	9587.14

The country's total installable wind power capacity is estimated to be of the order of 45,000 MW at 50 m above ground level. As against this figure, the actual installed capacity was only about 9600 MW at the end of 2008. Among the different states of India, Tamil Nadu's share in the total installed capacity is as high as 40 % and in fact it has exceeded its estimated potential. Tamil Nadu has the largest number of wind power installations. The states of Maharashtra, Gujarat, and Karnataka are also doing well, while other states like Andhra Pradesh and Madhya Pradesh still fall far short of the realizable capacity (Table 7.4.1).

As per the Indian Wind Energy Association (InWEA 2009) wind power currently accounts for 6 % of India's total installed power capacity, and it generates 1.6 % of the country's power. It is estimated by that by the year 2012 an additional wind power capacity of 6000 MW would have been installed in India. If the off-shore potential is also taken into consideration, India's total wind power potential would be as high as 65,000 MW. Therefore there is a tremendous scope for expansion of India's wind energy sector on the coming years.

India can be regarded as a new entrant into the field of wind power generation compared to countries like Denmark and the U. S. The systematic exploitation of wind energy began in India only twenty years ago but it has shown continuous and significant growth since then. This is the result of a combination of pragmatic, liberal and farsighted policies of the Government of India, which has established a separate Ministry for New and Renewable Energy, and the emergence of a strong private sector in this field.

According to the World Wind Energy Report 2008 (WWEA 2009), the total wind power installed capacity worldwide has now reached 121,000 MW of which 27,000 MW was added in 2008 alone. For the last several years, Germany, U. S., Spain, India and China have remained the top five countries in wind energy utilization, but their relative rankings have been varying. India's position has been either 4 or 5. For the first time in more than a decade, the U. S. has taken over the first position from Germany in terms of total installations. China continues its most dynamic role in the wind energy market and in the year 2008, it has more than doubled the installations for the third time in a row. China's growth rate in the wind energy sector now far outstrips that of India (Table 7.4.2). After carefully calculating and taking into account some insecurity factors, WWEA estimates that by 2020, wind energy will be able to have a share of at least 12% in the global electricity generation..

Table 7.4.2 Relative Ranking of Top Ten Wind Power Generating Countries of the World as of 2008 end (Source: WWEA 2009)

World Ranking as of 2008	Country	Total Capacity installed 2008 end (MW)	Capacity Added in 2008 (MW)	Growth Rate in 2008 (%)
1	U. S.	25170.0	8351.2	49.7
2	Germany	23902.8	1655.4	7.4
3	Spain	16740.3	1595.2	10.5
4	China	12210.0	6298.0	106.5
5	India	9587.0	1737.0	22.1
6	Italy	3736.0	1009.9	37.0
7	France	3404.0	949.0	38.7
8	U. K.	3287.9	898.9	37.6
9	Denmark	3160.0	35.0	1.1
10	Portugal	2862.0	732.0	34.4

Among the many impediments that have come in the way of commercial exploitation of wind power is its comparatively higher cost of production. However, as new technologies get developed in course of time and as the use of wind power proliferates further, the cost of production is bound to come down. In India, more than 95 % of installed wind power capacity belongs to the private sector which would help to keep the price competitive and affordable.

An important consideration that needs to be specially mentioned is that the use of wind energy is subject to seasonal and diurnal variations of the prevailing winds at any location and so the winds are not guaranteed to keep blowing all the time. As the power produced is by nature intermittent and is prone to disruption, a backup power source must be available to the users. In other words, electricity generated from wind power has to be supplied through a larger electricity grid that has alternative sources of power incorporated in it.

7.5 Food Security

At independence, India's population was 350 million. It crossed the one billion mark in the year 2000 and the census of India carried out in 2001 put the official figure at 1,028,737,436. What will India's future population be like is a matter of great interest to the world and concern for our country. Although the past and current increasing trends in India's population cannot just be extrapolated into the future, it is possible to make projections on the basis of trends in the fertility and mortality rates. Demographics is an evolving science, and depending upon what factors are assumed, different projections can be generated, somewhat like climate modelling.

The Population Foundation of India, New Delhi, and the Population Reference Bureau, Washington, DC, have together addressed the issue of India's population and attempted to project it across the next 100 years (Nanda and Haub 2007). They have offered two different scenarios of India's future population but both assume that fertility rate will decline continuously to the point where on an average a couple has two children, as per the goal of India's National Population Policy 2000. Scenario A assumes that the currently high fertility rates in some states of India will decline to 2.1 children, while Scenario B assumes that the decline will continue to 1.85 children, near the level observed in some states. Under Scenario A, India will cross the 2 billion population mark around 2071 and reach 2.181 billion at the beginning of the 22nd century. In Scenario B, the population peaks at 1.881 billion in 2081 and then begins to decline (Figure 7.5.1).

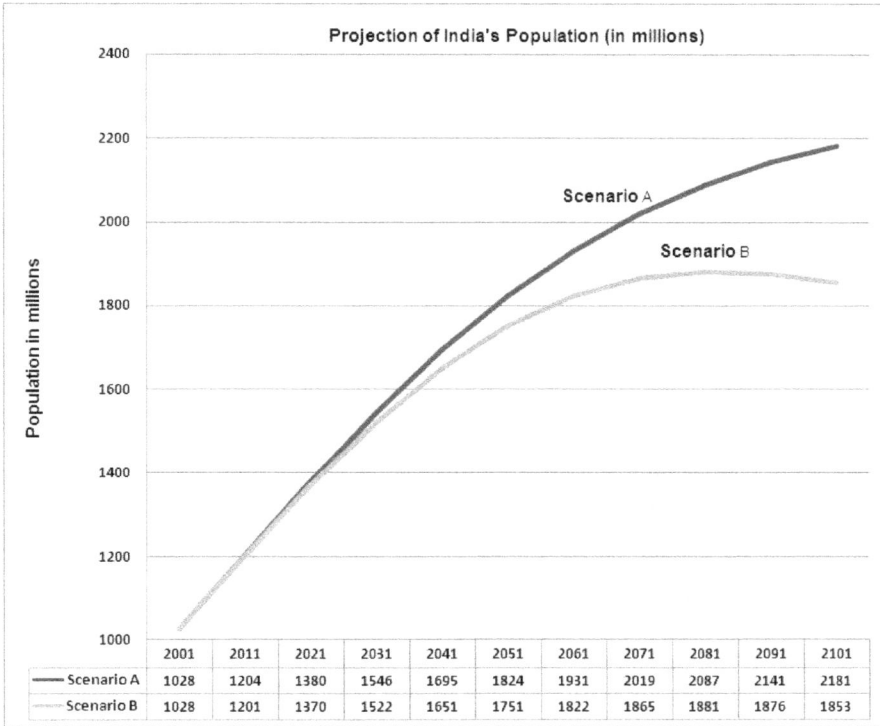

Figure 7.5.1 Projections of India's population into the 22nd century
(Data Source: Nanda et al 2007)

Between now and the middle of this century, a 60 % increase in population would mean a proportionate increase in the demand for food. However, additional land cannot be made available for cultivation at that rate and the area presently used for agriculture may even reduce as time goes by. Although the ever-increasing food demand can be met to some extent by exerting pressure on the agricultural system to increase production, there are many limiting considerations that cannot be overlooked. In fact it has been difficult to sustain the annual growth rates of food grain production and yield which were realized during the period of the green revolution. Many parts of the country in which the green revolution had shown the best results, are now showing signs of increasing environmental degradation such as declining soil fertility, falling water tables, increasing soil salinity and development of resistance to pesticides.

The M. S. Swaminathan Research Foundation has produced three atlases: the Food Insecurity Atlas of Rural India, the Food Insecurity Atlas of Urban India and the Atlas of the Sustainability of Food Security in India (MSSRF 2001, 2002, 2004). These publications reveal many important points that need to be addressed. In some states like Orissa, Himachal Pradesh, Bihar, Karnataka and Tamil Nadu, the net sown area has been declining. This process may lead to the shifting of prime agricultural land to non-agricultural uses. Land degradation has been fairly high in the states of Nagaland, Sikkim and Himachal Pradesh. There has been a dangerous overexploitation of ground water in Punjab, Haryana and Tamil Nadu. In states like Bihar, Uttar Pradesh and Madhya Pradesh, there is an urgent need to diversify livelihoods to non-crop and non-agricultural enterprises. Specifically, Punjab and Haryana which are the nation's chief grain suppliers are likely to reduce their production potential in a few decades if corrective measures are not taken against groundwater extraction, soil salinization and rice-wheat monoculture that are now prevalent there.

Aggarwal et al (2008) have carried out an exhaustive quantitative analysis of the yield gaps in four of India's major rainfed crops: rice, wheat, cotton and mustard. Presently rainfed agriculture in India covers an area of 94 million hectares, which has great potential for higher productivity. Theoretically, the potential yields estimated by Aggarwal et al (2008) using the InfoCrop model, are achievable by the current crop varieties with soil and weather as the only limiting factors. The biophysical yield gap is the difference between the estimated potential yield of a crop over a region and the region's actual average yield.

The results of Aggarwal et al have shown that in the case of the rice crop there is a considerable yield gap in all states, the country average being 1670 kg/ha. The yield gaps were the smallest in West Bengal and highest in Uttar Pradesh. However, in the case of wheat, the measured yields were significantly high due to irrigation and so the calculated yield gaps were negligibly small although in the states of Karnataka, West Bengal and Madhya Pradesh the yield gap is of the order of 80-800 kg/ha. The study of Aggarwal et al generally raises the hope that there is still a considerable scope for achieving higher yields in rainfed crops for meeting the increasing food requirements of the Indian population.

As the 21st century progresses, India will have to face a serious problem of food security, alleviating poverty, improving living standards and at the same time ensuring that this not done at the cost of the environment. Another aspect of the scenario of the future is the process of globalization that will call for structural changes in the agricultural sector if it is to remain globally competitive. These are matters that require attention and resolution irrespective of whether climate change is taking place or not. Climate change is just going to add one more dimension and an extra measure of uncertainty to a pre-existing problem which is already complex. This requires the adoption and implementation of a sound and far-reaching food security policy for India that would preserve the ecological foundations essential for sustainable food security besides giving an impetus to a sustainable intensification of agricultural production.

7.6 Ethics of Climate Change

Every living being requires a favourable environment not only to thrive but even just to exist. This is true for human beings as well, but the difference is that human beings have a far greater capability than all other creatures to overcome the constraints imposed upon their lives by weather, climate and other environmental factors. If the climate that people have got used to is changing, then it is a matter which directly affects them, both individually and collectively.

Climate change, however, is a very complex issue and it can be seen from several different angles. The science of climate change considers the observational evidence, attempts to model the processes within the climate system, and ventures to make projections and predictions of climate over the next century or longer. Climate change has a major political dimension, because when developed and developing countries come together to discuss it, the past history of exploitation comes into conflict with the hopes for a better future. Economics also comes into the picture as unabated climate change is sure to harm national economies, while mitigation actions have their own costs.

This book has so far attempted to take a rational, balanced and holistic view of different aspects of climate change. It has considered the basic science as well as the global politics and economics of climate change. It has discussed the likely future impacts on the Indian monsoon, India's agriculture, the vulnerability of its coastal zone, and the well-being of its people. A holistic view, however, must include a look at the ethics of climate change, and this is what the book is going to do in conclusion.

To be able to distinguish between right and wrong, good and bad, true and false, just and unjust, is a unique quality of the human mind. Ethical principles have always governed human societies and it is common these days to prescribe codes of conduct in different walks of life. Climate change is being universally discussed and every person is being urged to do something to save the planet. Therefore, the matter of climate change cannot remain an exception to the application of ethical considerations. In fact it needs it even more.

The first point to be considered is the ownership of the earth. The first verse of the Bible expresses the great story of God's creation in just ten words: "In the beginning, God created the heavens and the earth" (Genesis 1:1). Later on it reasserts: "The heavens declare the glory of God, the skies proclaim the work of his hands" (Psalm 19:1), and once again: "The earth is the Lord's and everything in it" (Psalm 24:1). There is no ambiguity about who owns this earth, it is God. The Bible is clear again about man's ownership rights, that he has none. "We brought nothing into the world, and we can take nothing out of it" (1 Timothy 6:6). "A man...as he comes so he departs, he takes nothing from his labour that he can carry in his hand" (Ecclesiastes 5:14).

The second point is about man's exploitation of natural resources. The Bible clarifies the relationship that God wanted to establish between man and nature: God wanted human beings to "...fill the earth and subdue it. Rule over the fish of the sea and the birds of the air and every living creature that moves on the ground....I give you every seed-bearing plant on the face of the whole earth and every tree that has fruit with seed in it..." (Genesis 1:28-29). Thus everything in nature was made freely available to man for use and enjoyment. The exploitation of nature by man has God's sanction.

The third point is about the consequences of man's actions. Agriculture is perhaps the most legitimate and inoffensive manner of exploitation of nature by man: "A man reaps what he sows" (Galatians 6:7). This law is not restricted to an agricultural process. It is a universal law found in many scriptures that the fruits of our actions eventually come back to us. However, it also happens that "while one sows, another reaps" (Ecclesiastes 9:11). It is clear that we have sown carbon dioxide in the atmosphere, and we are now reaping the harvest of global warming. It is the western industrialized nations that have sown carbon dioxide, but it is the poorer developing nations who are reaping the ill-effects through the unified climate system of the earth.

Even if one has a different set of beliefs, no human being can possibly stake a claim to the ownership of the earth. Even what we legally own, be it land, material wealth or intellectual property, is ours only in a temporary and relative sense. One of the most famous short stories of Leo Tolstoy had as its title this question: "How much land does a man need?" The answer provided at the end of the story was "six by three", signifying that a plot of that size would be enough to bury not just a man's body but also his ambitions and greed.

There are two parables of Jesus (Matthew 21:33-44, 25:14-30) about a master who has to go away leaving his property in charge of servants. The master expects his trusted servants to take care of the property and put it to good use in his absence, but that does not happen. These parables are equally applicable to man's use of the environment. Man is still free to use all that nature provides and that includes land, oceans and the atmosphere. Nature does not ask for a payment in return for oxygen, water or sunlight which are essential for our remaining alive on earth. However, with this great power to exploit nature, comes an equally great responsibility. When we get something free, we have a choice: we can either be careless and destroy it, or we can be caring and nurture it.

Protection and conservation of nature are necessary in order to maintain the world's ecological balance. However, there is no need of going to the extent of regarding nature as something sacred and holding human beings responsible for actions against nature. It is dangerous to develop a feeling of guilt in people's minds and calling for an atonement. It makes little sense to ask people in India to conserve electricity in the name of climate change when its smaller villages have no supply of electricity and the biggest cities face load shedding for several hours a day regularly. And that too while the developed countries continue to increase their wasteful power consumption without any remorse. It is wrong to tell our schoolchildren to study in candlelight while lights are switched off to commemorate earth day. Such symbolism takes many forms such as asking people to join in a marathon race wearing t-shirts flaunting slogans like 'save the planet', or conducting meetings on mountain tops or under the ocean. Such symbolism is good for creating awareness but it cannot reduce global warming.

Yes, we have a responsibility towards the coming generations. We should not misuse the environment. We should leave the earth a better place than we found it to be. But for doing this we need not go back to the dark ages.

7.7 References

Aggarwal P. K. and coauthors, 2008, "Quantification of yield gaps in rain-fed rice, wheat, cotton and mustard in India. Global theme on agroecosystems", *Report No. 43*, International Crops Research Institute for the Semi-Arid Tropics. Patancheru, India, 36 pp.

IMD, 2009, *Solar Radiant Energy over India*, India Meteorological Department, 876 pp.

Mani A., 1981, *Handbook of Solar Radiation Data for India,* Allied Publishers Pvt. Ltd., New Delhi.

Mani A. and Rangarajan S., 1982, *Solar Radiation over India,* Allied Publishers Pvt Ltd., New Delhi.

InWEA, 2009, Indian Wind Energy Association web site http://www.inwea.org

MoEF, 2009a, *National Action Plan on Climate Change,* Ministry of Environment and Forests, Government of India, 49 pp.

MoEF, 2009b, *India and Climate Change,* Ministry of Environment and Forests, Government of India, 24 pp.

MSSRF, 2001, *Food Insecurity Atlas of Rural India,* M. S. Swaminathan Research Foundation, Chennai.

MSSRF, 2002, *Food Insecurity Atlas of Urban India,* M. S. Swaminathan Research Foundation, Chennai.

MSSRF, 2004, *Atlas of the Sustainability of Food Security in India,* M. S. Swaminathan Research Foundation, Chennai.

Nanda A. R. and Haub C., 2007, *The future population of India - A long-range demographic view,* Population Foundation of India, New Delhi, and Population Reference Bureau, Washington, DC, USA, 20 pp.

WWEA, 2009, *World Wind Energy Report 2008,* World Wind Energy Association, Bonn, Germany, 16 pp.

Subject Index